AF872911

Rigid-Flexible Coupling Dynamics and Control of Flexible Spacecraft with Time-Varying Parameters

Jie Wang · Dong-Xu Li

Rigid-Flexible Coupling Dynamics and Control of Flexible Spacecraft with Time-Varying Parameters

Springer

Jie Wang
National University of Defense Technology
Changsha, Hunan, China

Dong-Xu Li
National University of Defense Technology
Changsha, Hunan, China

ISBN 978-981-16-5096-3 ISBN 978-981-16-5097-0 (eBook)
https://doi.org/10.1007/978-981-16-5097-0

© The Editor(s) (if applicable) and The Author(s), under exclusive license to Springer Nature Singapore Pte Ltd. 2022
This work is subject to copyright. All rights are solely and exclusively licensed by the Publisher, whether the whole or part of the material is concerned, specifically the rights of translation, reprinting, reuse of illustrations, recitation, broadcasting, reproduction on microfilms or in any other physical way, and transmission or information storage and retrieval, electronic adaptation, computer software, or by similar or dissimilar methodology now known or hereafter developed.
The use of general descriptive names, registered names, trademarks, service marks, etc. in this publication does not imply, even in the absence of a specific statement, that such names are exempt from the relevant protective laws and regulations and therefore free for general use.
The publisher, the authors and the editors are safe to assume that the advice and information in this book are believed to be true and accurate at the date of publication. Neither the publisher nor the authors or the editors give a warranty, expressed or implied, with respect to the material contained herein or for any errors or omissions that may have been made. The publisher remains neutral with regard to jurisdictional claims in published maps and institutional affiliations.

This Springer imprint is published by the registered company Springer Nature Singapore Pte Ltd.
The registered company address is: 152 Beach Road, #21-01/04 Gateway East, Singapore 189721, Singapore

Preface

Large flexible appendages, such as solar wings and deployable antennas, usually rotate relative to the platform of the spacecraft. And the relative motion would result a nonlinear system with time-varying parameters. At the same time, the coupling of the elastic vibration of the flexible structure with the orbit and attitude movement of the spacecraft platform brings severe challenges to the spacecraft control.

The subject of this book is the coupling dynamics and control of flexible spacecraft with time-varying parameters. The dynamic characteristics, vibration control methods and attitude stabilization methods for spacecraft, are systematically studied in respects of the theoretical modeling, numerical simulation, and the ground experiment. The first-order coupled dynamic models for flexible spacecraft with time-varying configuration are established. Characteristics of the appendages and effects on the spacecraft platform are analyzed based on the complex mode. Three active control theories, i.e., linear state feedback, variable positive position feedback, and sliding mode control methods are proposed for systems in complex mode space. In order to solve the problem of high-precision attitude control for flexible spacecraft with time-varying parameters, an optimal variable amplitudes input shaping control method, an adaptive sliding mode control method, and an attitude coupling control method are proposed for slew maneuver of the flexible spacecraft. The flexible spacecraft testbed is established to validate the proposed controllers. Ground tests of attitude maneuver and stabilization are conducted to validate the optimal variable amplitudes input shaping control method, the adaptive sliding mode method, and the coupling control method.

Chapter 1 provides an introduction to the concepts presented in the text. Chapter 2 provides the mathematic models for flexible spacecraft. Chapter 3 carries out the model analysis work for the model established in the second chapter. Chapter 4 presents methods and techniques for vibration control of flexible structures in complex space. Chapter 5 presents kinds of slew maneuver trajectories based on the variable amplitudes input shaping methods for flexible spacecraft. Chapter 6 proposes coupling control methods based on robust control, positive position feedback and the sliding mode control methods. Chapter 7 establishes an attitude control testbed for flexible spacecraft and carries out attitude maneuver experiments and coupling control experiments, respectively. Chapter 8 is the prospects.

In writing this book the authors hope to make a contribution to the design and analysis of flexible spacecraft. The research provides an important way to solve the problem of high-precision attitude control of flexible spacecraft with time-varying parameters. This text addresses key issues of flexible spacecraft with time-varying parameters. And the analyses are meaningful for spacecraft to achieve high precision.

Changsha, China

Jie Wang
Dong-Xu Li

Contents

Chapter 1
Introduction

Since the launch of the first satellite-Sputnik 1 in 1957, in only sixty years, the space industry has written a glorious chapter in the long history. Countries around the world have launched thousands of spacecraft. Some of them are operating in low-Earth orbit to serve the day-to-day life of humanity. The others travel through the vast and deep space to explore the mysteries of the universe. No matter what mission the spacecraft carries out, attitude control is an important part of the spacecraft to complete the mission. In recent years, with significant improvements in science and technology and an urgent increase in demand, some large-scale spacecraft have tended to be more flexible. At the same time, the spacecraft attitude control has to meet the demand of high precision and high stability. As a result, the design of the spacecraft attitude control system has been seriously challenged.

This article focuses on the rigid-flexible coupling dynamics and control of modern large-scale flexible spacecraft. In this chapter, the background to the research is clarified, and the scientific meaning and application value of the research is described in more detail. Then it summarizes the state of the research and presents the key points and difficulties of the research. Finally, the major research content, research ideas and organizational structure of the book are briefly described.

1.1 Development Trends for Modern Spacecraft

Large modern spacecraft (Figs. 1.1 and 1.2), such as communication satellites, Earth observation satellites, and manned spacecraft, are mainly characterized by a rigid central body with one or more large-scale flexible attachments. The size of the latter after unfolding is generally much larger than the central rigid body. The elastic deformation of the flexible attachments must be taken into account in the analysis. This type of structure is referred to as flexible structure [1], and the type of spacecraft with

© The Author(s), under exclusive license to Springer Nature Singapore Pte Ltd. 2022

J. Wang and D.-X. Li, *Rigid-Flexible Coupling Dynamics and Control of Flexible Spacecraft with Time-Varying Parameters*,
https://doi.org/10.1007/978-981-16-5097-0_1

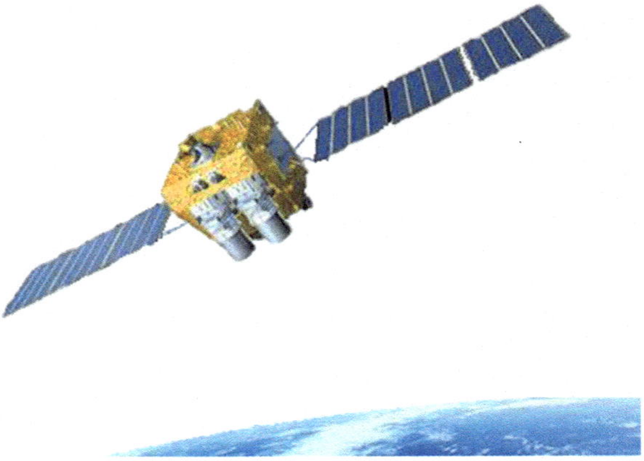

Fig. 1.1 GF-2 earth observation satellite

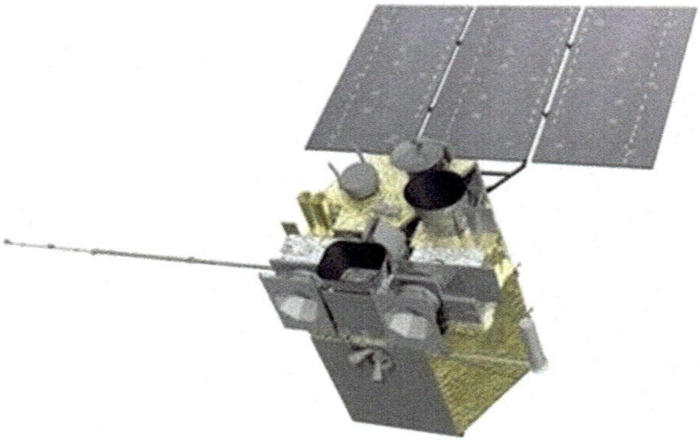

Fig. 1.2 FY-4 meteorological satellite

flexible structure is defined as a flexible spacecraft [2]. Take Earth observation satellites as an example, in addition to employing large solar wings, this type of satellite also carries payloads such as cameras, space interferometers, or large-scale synthetic aperture radar antennas (SAR antennas). Thus, it is typically a flexible spacecraft. These spacecrafts are widely used in various fields of aerospace engineering and are of great importance to the improving human life.

With the development of the aerospace engineering, some spacecraft have shown a trend of high precision, high agility, large-scale, and high flexibility.

High precision: Take the earth observation satellite as an example. In order to make the details of the observation target clearer, the resolution of the satellite payload is required to be higher. And the satellite must have high pointing accuracy and attitude stability. Due to the improved resolution and positioning accuracy, the requirements for satellite pointing accuracy and stability are listed in Table 1.1 [3]. The ESA ENVISAT satellite requires a pointing accuracy of 0.035 and a pointing stability of $3 \times 10^{-3\circ}$/s [4]. The attitude stability of Japan's Advanced Land Observing Satellite (ALOS) is $2 \times 10^{-4\circ}$/5 s [5].

Table 1.1 Pointing accuracy and stability of earth observation satellites

era	Representative satellite	Pointing accuracy /(°)	Attitude stability /(°/s)
1970s	Seasat\Landsat-1	1 ~ 0.3	$5 \times 10^{-2} \sim 1 \times 10^{-2}$
1980s	SPOT\Landsat-2	0.3 ~ 0.03	$3 \times 10^{-3} \sim 3 \times 10^{-5}$
1990s	ADEOS\JERS	0.3 ~ 0.02	$1 \times 10^{-3} \sim 1 \times 10^{-6}$
2000s	Hilios-2\IRS-P	0.1 ~ 0.001	Better than 1×10^{-4}

High agility: Agility is one of the major development trends for flexible spacecraft. Compared to the traditional spacecraft, the time for a satellite with fast maneuverability to get the same amount of information is significantly shortened, so its efficiency is significantly improved. In a series of Earth observation programs in Europe and the US, the fast maneuverability of satellites has been put forward. Some of the representative satellites are Orbview-5, Pleiades, IKonos-2, WorldView, Quickbird, etc. For example, Orbview-5 has the maneuverability of ±60°. The WorldView-1 satellite has the maneuverability of ± 40°. The maximum attitude maneuvering speed can reach the value of 4.5°/s, and the maximum maneuvering angular acceleration can reach 2.5°/s^2 [6].

Large-scale: To meet the needs of diverse missions and payloads, the large scale is one of the main development trends for flexible spacecraft. For the solar wing, in order to supply enough energy for satellite, the solar wing area showed a steady upward trend. Limited by the conversion efficiency of solar cells, the area-to-power ratio of the solar wing is around 5 m^2/kW [7]. In order to provide several kilowatts or even ten kilowatts of power, the area of the solar array can reach tens of square meters. For example, the size of the solar wing of the ALOS is 3 × 22 m [8]. The platform of the most advanced communication satellite requires a power of no less than 15 kW at the beginning of the life, and the area of solar array can reach nearly 100 m^2. For deployable spaceborne antennas, increasing aperture increases the intensity of reflected signals, improves beam pattern properties, and increase resolution. The size of the C-band SAR antenna mounted on the Canadian RADARSAT-2 satellite is 15.0 × 1.5 m [9], and the size of the C-band advanced synthetic aperture antenna (ASAR) mounted on the ENVISAT satellite is 10.0 × 1.3 m [10]. The L-band phased array synthetic aperture antenna (PALSAR) carried by ALOS has a size of 8.9 × 3.1 m [11].

Flexibility: Limited by the cost of space launches, light weight has become an inevitable choice for large-scale structures in satellites, which has led to a trend toward flexibility in structures such as solar wings and antennas. In order to improve the power-to-mass ratio, the solar array has undergone a steady evolution from rigid panel planar arrays to flexible folding arrays or roll-out arrays. The power-to-mass ratio of the traditional rigid folding solar panel is about 45 W/kg, and for the flexible folding solar wing the ratio is raised to 60 W/kg [12]. The flexible solar wing is lighter to provide the same power. Reflector antennas were commonly used in early spacecraft, with an antenna mass-to-area ratio of approximately 3 kg/m^2. In order to reduce the structural mass, the large cable net antennas have been developed. The diameters of some antennas have reached several tens of meters, and the mass-to-area ratio has reached 1 kg/m^2 [13]. Due to the adoption of lightweight configurations or materials, the flexibility of the system increases and the natural frequency decreases.

1.2 Flexible Spacecraft with Time-Varying Parameters

With the complexity and diversity of space missions, there have single or multiple flexible rotatable attachments in some spacecraft, which experience a wide range of rigid body motions relative to the spacecraft body due to needs of the missions. For instance, when the spacecraft is orbiting, the solar wing needs to keep track of the sun; when performing the attitude maneuvering task, the solar wing needs to be quickly oriented and stabilized after the maneuvering process. In some cases, the antenna must also continually orient toward the target. For example, the L-band reflector on the NASA's 2015 Soil Moisture Active Passive (SMAP) satellite rotates around the zenith axis at a rate of 14.6 rpm (87.6°/s) for a wide range of observations. It is necessary to ensure the pointing accuracy of the antenna surface for a high-precision remote mission of Earth [14, 15].

The rigid motion of a flexible structure causes the spacecraft to exhibit significant characteristics of time-varying parameters. When the structure such as a solar wing or an antenna rotates relative to the spacecraft body, the dynamical parameters of the whole system are time-varying. For example, the rotation of the solar wings in Japan's Engineering Test Satellite VIII (ETS-VIII) around the pitch axis can cause a variation of $\pm 25\%$ of the system dynamical parameters [16]. Changes in kinetic parameters will also cause changes in the dynamics of the system. In the ETS-VIII satellite, the rotation of the solar wings causes the natural frequency variation of the satellite to be up to 12% [17]. Also, in one cycle of antenna rotation, the first-order flexible modal frequency of the SMAP satellite varies by 10% [18]. Due to the time-varying characters of the kinetic parameters, the dynamic analysis and control of the flexible spacecraft poses a huge challenge for scholars [19].

1.3 Problems

(1) Vibration of large flexible structures.

Large-scale flexible structures (solar wings, antennas, etc.) are inevitably affected by various external and internal forces in space. Due to their low rigidities, they can be easily stimulated. In addition, the atmospheric damping in the space environment is weak. Once the vibration is excited, it takes a long time to attenuate, which not only affects the strength and stability of the flexible structure, but also reduces the accuracy of the spacecraft, and even leads to the failure of space missions. For example, the thermal loads in the space influence the flexible structure in the entire life of a spacecraft. The heat source inside the spacecraft, the radiation absorbed by the spacecraft from external heat sources (solar radiation, planetary radiation and solar radiation reflected by the planets), and the spacecraft's deep-space radiation, etc. heat the spacecraft structure unevenly. And as the spacecraft alternately operates in the sun's irradiated area and the earth's shadow area, the structure of the spacecraft is in a temperature field with time-varying values for a long time. This will cause thermally induced vibrations of the structure, thereby reducing the accuracy of the spacecraft.

(2) Large-angle attitude maneuver of spacecraft.

In the process of large-angle attitude maneuver, the modal vibration of flexible structures is easy to be stimulated. After the elastic vibration of the flexible structure to be excited is completely attenuated and the attitude accuracy of the spacecraft body meets the requirements, the payload starts to work. However, it is difficult to rapidly attenuate the elastic vibration of the flexible structure by relying solely on the thin atmosphere and the weak damping of the material. On the other hand, the flexible spacecraft performing attitude maneuver is a nonlinear time-varying parameter system. When the satellite is performing a small angle maneuvering, the nonlinear term in the attitude control equation can be neglected. While for a large-angle attitude maneuver, the nonlinear term cannot be neglected. Thus, it is necessary to consider the characteristics of the system when designing the satellite attitude control system.

(3) High-precision attitude control of spacecraft.

When designing the attitude control system of a spacecraft according to the traditional method, it is necessary to obtain the nominal dynamic parameters of the system. However, the accuracy of these nominal parameters cannot be guaranteed. On one hand, due to fuel consumption and vibration of flexible attachments, there are deviations in the estimation of mass, inertia and other parameters of the spacecraft. The performance of sensors and actuators will also change with respect to time. When the large-scale flexible solar wing or antenna has rigid body motion relative to the spacecraft body, it will cause obvious changes in dynamic parameters such as the system inertia and rigid-flexible coupling coefficient. On the other hand, modern spacecraft is intended to have higher performance, especially for attitude pointing accuracy and stability. Controllers based on nominal parameters cannot meet the requirements of missions. In addition, due to the trend of large-scale and flexibility, the natural frequency of the structure is reduced, so that the band of the vibration and the band of the attitude control system are coupled with each other. All these have brought difficulties to the design of satellite attitude control system.

1.4 Aims and Outline of the Book

This book focuses on solving the following three problems:

The first one is the rigid-flexible coupling dynamic modeling and analysis of flexible spacecraft, which belongs to the category of dynamic modeling of flexible multi-body systems. The coupled dynamic equations of orbit and attitude motions of the spacecraft and the elastic vibration of flexible structures have strong nonlinear characteristics, and the relative motion between the platform and the attachment induces the time-varying dynamic parameters. Therefore, it is of great significance to study the dynamic modeling problem of the nonlinear system with time-varying parameters and to consider the influence of the relative motion on the system characteristics, which will lay the foundation for the flexible spacecraft attitude control and flexible structure control.

The second is the vibration control of flexible appendages in the spacecraft, which belongs to the category of structural vibration control. After the flexible appendage is disturbed, its elastic vibration should be attenuated as soon as possible to reduce the impact on the pointing accuracy and stability of the spacecraft platform. On the one hand, large-scale flexible structures in space are usually combinations of rods, beams, plates, shells, and bodies, which have complex dynamics. Both the model uncertainty brought by modeling and the saturated nonlinearity of the actuator brings challenges to the design of controllers. On the other hand, when the flexible appendage has rigid motion relative to the platform, especially when it is rotating, the gyro term exists in the governing equation. As a result, the equations should be decoupled and reduced in complex space instead of real space. Therefore, designing controller in complex space is an important issue for the research of structural vibration control.

The third is the attitude maneuvering and stability of flexible spacecraft, which belong to the category of spacecraft attitude control. Compared with classical attitude maneuvering methods, a maneuver strategy is proposed to ensure the elastic vibration of the flexible appendage as small as possible, which will greatly shorten the maneuvering time. At the same time, studying the coupling control method of attitude motion and elastic vibration will further improve the attitude accuracy and attitude stability of the spacecraft platform, which is of great value for the effectiveness of the spacecraft.

Taking the flexible spacecraft with time-varying parameters as the research object, the book develops rigid-flexible coupling dynamic modeling and control methods for the system with time-varying parameters, establishes the dynamic model for the whole system, and proposes vibration control methods and spacecraft attitude control methods to provide theoretical support for the research of modern spacecraft. The full text is divided into eight chapters:

This chapter is introduction.

Chapter 2 completes the modeling of the research object. First, the dynamic characterization of the flexible spacecraft with time-varying parameters was introduced, and the simplified models of the accessories and the whole system were given. Then, the flexible accessories and the whole system of the spacecraft were established in turn. Rigid-flexible coupling model. When modeling the flexible attachment, the main motion effect of the piezoelectric material was considered, and the electromechanical coupling model of the flexible attachment was established.

Chapter 3 carries out the model analysis work for the model established in the second chapter, including the dynamic characteristics analysis of the flexible attachment in the case of rigid body motion, and the analysis of the dynamic characteristics and response effects of variable parameters on the rigid-flexible coupling system of the spacecraft.

Chapter 4 presents methods and techniques for vibration control of flexible structures in complex space. Based on the complex modal theory, the design methods of the state feedback controller, the positive position feedback controller and the sliding mode controller are respectively proposed, and numerical simulations are carried out.

In Chap. 5, based on the coupled dynamics model of the flexible spacecraft established in Chap. 2, the attitude control method is studied based on the input shaping method. First, based on the traditional input shaping method, the attitude maneuver path of the fixed parameter system is designed. Then, for the flexible spacecraft with variable parameters, a method of forming attitude maneuvers with variable amplitude input is proposed. Finally, in order to improve the robustness, a robust attitude maneuvering method for variable amplitude input shaping is proposed.

Chapter 6 proposes robust H_∞ and adaptive sliding mode attitude control methods for flexible spacecraft with variable parameters, as well as coupling control methods based on robust control and positive position feedback, sliding mode control and positive position feedback, and carries out numerical simulation research.

Chapter 7 builds the flexible spacecraft coupling control experiment system, carries out the attitude maneuver experiment and the coupling control experiment, respectively used to verify the attitude maneuver method based on input shaping proposed in Chap. 5 and the adaptive control method proposed in chap. 6 Sliding mode control method and coupling control method.

Chapter 8 is the prospects.

References

1. Zhang, R. (1998). *Satellite orbit attitude dynamics and control [M]*. Beijing University of Aeronautics and Astronautics Press.
2. Li, D. (2010). *Structural dynamics of flexible spacecraft [M]*. Science Press.
3. Shi, S. (2000). Study on high precision attitude control systems o fforeign earth-observation satellites. *Aerospace Shanghai, 17*(6), 49–53.
4. Bargellini, P., Matatoros, M. A. G., Ventimiglia, L., et al. (2005) ENVISAT attitude and orbit control in-orbit performance: an operational view. In *Proceedings of the 6th International ESA Conference on Guidance, Navigation and Control Systems*. Loutraki: Greece.
5. Kimura, H., Ito, N. (2000) ALOS/PALSAR: The Japanese second-generation spaceborne SAR and its applications. In *Proceedings of the SPIE*.
6. Ye, D. (2013) Research on fast maneuver and stabilization control algorithm for agile satellite. Harbin Institute of Technology
7. Luque, A., & Hegedus, S. (2003). Handbook of photovoltaic science and engineering. Hoboken, NJ: John Wiley & Sons Ltd
8. Iwata, T. (2008) Attitude dynamics and disturbances of the advanced land observing satellite (ALOS): modeling, identification, and mitigation. In *AIAA/AAS Astrodynamics Specialist Conference and Exhibit* (pp. 1–20). Honolulu, Hawaii.

9. Livingstone, C. E., Sikaneta, I., Gierull, C., et al. (2006) RADARSAT-2 system and mode description.
10. Desnos, Y-L., Buck, C., Guijarro, J., et al. (2000) ASAR-Envisat's advanced synthetic aperture radar. 91–100.
11. Ito, N., Hamazaki, T., Tomioka, K. ALOS/PALSAR characteristics and status.
12. Jones, P. A., & Spence, B. R. (1998) Spacecraft solar array technology trends. In Proceedings of the aerospace conference.
13. Imbriale, W. A., Gao, S. S., Boccia, L. (2012). Space antenna handbook. Hoboken, NJ: John Wiley & Sons.
14. Entekhabi, D., Njoku, E. G., O'Neill, P. E., et al. (2010). The soilmoisture active passive (SMAP)mission. *Proceedings of the IEEE, 98*(5), 704–716.
15. Liu, J. Y. (2014) Space-based large spinning sensor pointing and control design and its application to NASA's SMAP spacecraft. In *AIAA Guidance, Navigation, and Control Conference National Harbor* (pp. 1–18). Maryland.
16. Nagashio, T., Kida, T., Ohtani, T., et al. (2010). Design and implementation of robust symmetric attitude controller for ETS-VIII spacecraft. *Control Engineering Practice, 18*, 1440–1451.
17. Ni, Z., Tan, S., & Wu, Z. (2016). Identification of time-varying frequencies and model parameters for large flexible on-orbit satellites using a recursive algorithm. *Trans Japan Soc Aero Space Sci, 59*(3), 150–160.
18. Ni, Z., Mu, R., Xun, G., et al. (2016). Time-varying modal parameters identification of a spacecraft with rotating flexible appendage by recursive algorithm. *Acta Astronautica, 118*, 49–61.
19. Li, Z., Li, Y., & Li, G. (2005). Issue on high performance fast maneuver of variable parameter spacecraft. *Journal of Central South University, 36*(1), 8–13.

Chapter 2
Rigid-Flexible Coupling Dynamic Modeling of Flexible Spacecraft

2.1 Introduction

A modern spacecraft that typically includes flexible appendages, such as antennas, solar arrays, etc., is a very complex multi-body system. The flexibilities of the large appendages will affect the attitude of the entire spacecraft. As a result, the dynamics and control of flexible spacecraft have caught the attention of researchers over the last three decades. In many applications, flexible appendages need to be redirected to accomplish specific tasks. For example, after the attitude of the observation satellite is maneuvered, the antenna needs to be rotated to ensure a certain direction. The solar wing continues to rotate to face the sun to ensure maximum power. During the redirection of the appendages, the flexible spacecraft is a system that varies according to time. Therefore, a precise kinetic model of a flexible spacecraft with a time-dependent configuration is necessary to successfully predict the attitude and designation of the attitude and orbit control system (AOCS).

Quite often, flexible spacecraft was deemed to be a collection of interconnected rigid bodies to some of which are attached one or several flexible appendages [1]. The dynamic modeling theory of flexible multi-body system is developed in the theory of multi-rigid-body system dynamics [2]. After nearly forty years of development, considerable progress has been made. For the purpose of describing of the motion of the rigid-flexible coupling system, researchers have proposed several methods: the floating frame of reference, the convected coordinate system, finite segment method, the large rotation vector and absolute nodal coordinate formulation [3]. These methods were widely used in multi-body systems in various branches of engineering. Among these, the floating frame of reference formulation is the most widely used methods for flexible spacecraft. The governing equations for the system are mostly derived either by the Eulerian approach or the Lagrangian approach [4]. Meirovitch [5] proposed Lagrange's equations of motion in terms of

© The Author(s), under exclusive license to Springer Nature Singapore Pte Ltd. 2022
J. Wang and D.-X. Li, *Rigid-Flexible Coupling Dynamics and Control of Flexible Spacecraft with Time-Varying Parameters*,
https://doi.org/10.1007/978-981-16-5097-0_2

quasi-coordinates for a hybrid system. Then the method was used to derive general motions of flexible spacecraft with fixed appendages [6] and flexible multi-body systems [7].

First, the flexible spacecraft was idealized as a hub-beam system undergoes planar motion. So the coupling between the rotational motion of the hub and the bending deformation of the beam was analyzed. Then the system dynamics were extended to a hub-beams system with three-dimensional large overall deformation [8]. Liu and Lu [9] considered the effects of the torsional deformation as well as the longitudinal deformation of the beam on the overall motion of the hub. Some other literature focused on the effect of the geometrically nonlinear kinematics of deformation [10]. Gasbarri et al. [11] evaluated the influence of elastic vibrations on the system's inertia and investigated the contributions on the attitude control of the whole spacecraft. Based on the established model, lots of literature focused on the slew maneuver control design or attitude stabilization of the three-axis rotational flexible spacecraft. This method leads to equations of motion expressed in terms of a combination of discrete coordinates describing the arbitrary rotational motions of the rigid bodies and distributed or modal coordinates describing the small, time-varying deformations of the appendages [12].

Most literature dealt with dynamic and control of spacecraft with fixed appendages [13–15]. Only few literature focused on the dynamic model and control with rotating appendages [16]. In the case of rotating flexible structures, the floating frames must translate and rotate with the deformed structure that the structural displacements fall within the range of validity of the small deformation assumption [17, 18]. The transformation between the floating frame and the body frame is time-varying when the rotation of appendages is considered. The conventional model neglects the partial derivations of the matrix and would induce severe errors in a problem with the rotation of appendages. Therefore, the dynamic modeling and analysis of this type of flexible spacecraft needs further research.

The main purpose of this chapter is to establish a rigid-flexible coupled dynamics model of the flexible spacecraft with rotating structures. This chapter first analyzes the physical configuration of a typical flexible spacecraft. The motion description method is presented for flexible spacecraft with a rigid-flexible topology. And then simplified models for appendages and the whole spacecraft are established. For the flexible appendage with rigid motions, a dynamic model is established. Finally, for the flexible spacecraft with rotating appendages, a coupled dynamic model of the rigid motion with six degrees of freedom and the elastic vibration of the flexible appendage is established.

2.2 Description of Flexible Spacecraft with Rotating Appendages

2.2.1 System Description

Figure 2.1 presents a typical flexible spacecraft, which consists of a central core and several flexible appendages, for instance, solar wing and antenna. On one hand, the spacecraft platform experiences rigid body motion with six degrees of freedom in orbit. For example, when performing tasks such as remote sensing and communication, the spacecraft body needs to perform orbit or attitude maneuvers to aim at the target. On the other hand, in order to track the sun or the target on the ground, flexible appendages on the spacecraft (such as solar wings, satellite antennas) rotate relative to the spacecraft platform. Due to the coupling between the spacecraft platform and flexible appendages, the attitude of the platform will stimulate the elastic vibration of the appendages. In addition, some environmental disturbances, such as temperature and space debris collision, will also stimulate the elastic vibration of the flexible appendages, and the elastic vibration will affect the attitude stability and pointing accuracy of the platform.

According to the configuration of the flexible spacecraft, the spacecraft is idealized as a central body with one or more flexible appendages, as shown in Fig. 2.2. The spacecraft platform is simplified as a rigid body, and the appendages are simplified as flexible bodies. The flexible body is connected to the platform through connectors,

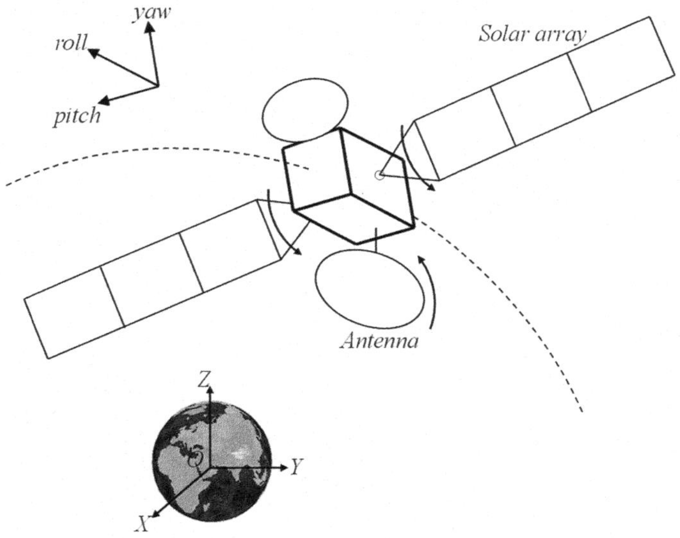

Fig. 2.1 Scheme of the flexible spacecraft

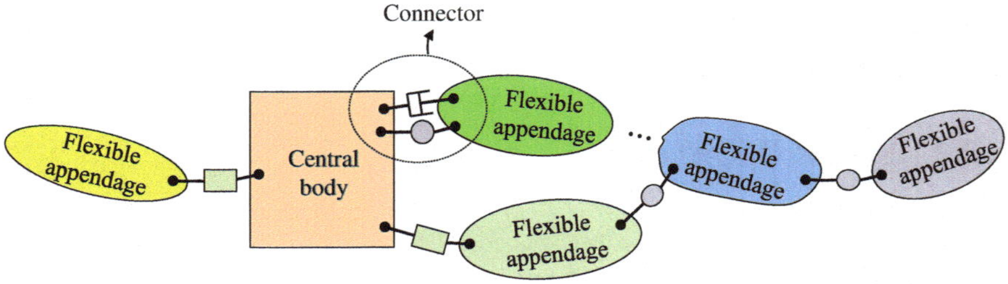

Fig. 2.2 Topology configuration of flexible spacecraft

which represent fixed connections, rigid hinges, flexible hinges, motors, or other damping structures.

2.2.2 Description of Motion

In order to formulate the equations of motion, the authors first introduce coordinate systems, as shown in Fig. 2.3.

(1) Frame *OXYZ* is an inertial frame which has an origin fixed at point *O*.
(2) Frame *oxyz* is a body frame, which coincides with the principal axes of the platform and with an origin at the mass center *o* of the platform. Since the platform is simplified as a rigid body, the motion of the satellite can be described by the translations and rotations of the *oxyz* coordinate system relative to the *OXYZ* coordinate system.
(3) Frame $o_s x_s y_s z_s$ is a floating frame, which is bound to the elastic appendage in the undeformed state and with the origin at the hinge point o_s. This coordinate system is used to describe the rigid motion of the flexible body relative to the inertial space, including translations and rotations. The origin is generally located at the connection between the flexible body and the spacecraft platform. The elastic deformation of the flexible body is described by the relative position

Fig. 2.3 Motion description of flexible spacecraft based on hybrid coordinate method

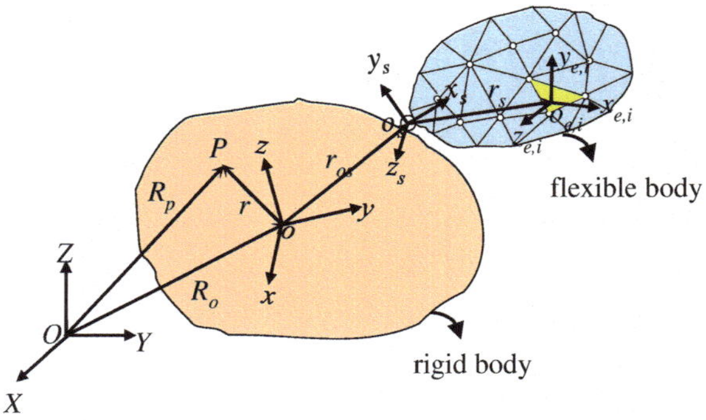

of the flexible body and its body coordinate system, and the flexible body can be discrete by the hypothetical mode method or the finite element method.

(4) Frames $o_{e,i}x_{e,i}y_{e,i}z_{e,i}$ ($i = 1,2,\ldots,N$) is the local coordinate of the i-th element of the appendage, whose origin is at the mass center $o_{e,i}N$ denotes the total number of elements.

2.2.3 *Model Simplification*

The spacecraft studied in this chapter focuses on the spacecraft whose flexible appendages such as solar wing and antenna rotate relative to the spacecraft platform. The simplified models of rotating flexible appendages and the whole system of flexible spacecraft are presented.

2.2.3.1 Simplified Model of Spinning Appendage

There are two forms of motion for rotating structures. One is rotations relative to the spacecraft platform. The other is the elastic vibration which is excited by the platform's motion or the environment. According to the configuration and movement form of the solar wing, antenna and other structures, it is simplified to a combined structure model of rotating flexible beam and beam plate.

In addition, due to the errors in the manufacturing process and the unbalanced distribution of solar cells, flexible cables and other components, the eccentricity of the structure is also considered in the simplified model.

(1) **Spinning flexible beam**

For the solar wing, antennas and other structures whose size in one direction is much larger than the size in the other two directions (such as the solar wing of ETS-VIII satellite with a size of 19 × 2 m), the flexible appendage is simplified to be a spinning beam with arbitrary cross-section, as shown in Fig. 2.4. The boundary condition is that the root is fixed, only the rotational freedom around the longitudinal axis is released, and the tip is free.

Taking the solar wing as an example, the basic assumptions for model simplification are as follows:

Fig. 2.4 Flexible spinning beam model **(b)** simplified model

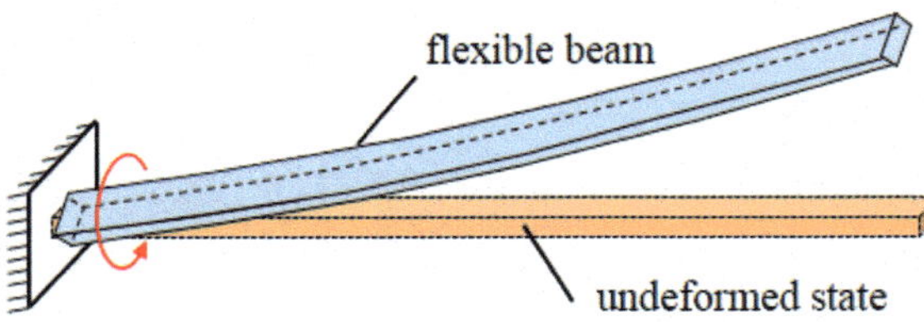

(b) simplified model

(a) Ignore the influence of the connecting structure (such as motor, slip ring) between the solar wing, antenna, etc. and the spacecraft platform. After the structure is unfolded and locked, the root can only rotate around the longitudinal axis, and the degrees of freedom in the other five directions are constrained.
(b) Only the influence of rotation on the solar wing is considered, and the influence of rigid motions in other directions of the spacecraft platform on the rigidity of the solar wing is ignored.
(c) Ignore the influence of the rigidity of the hinge connection between the yoke and the solar panel. Ignore the influence of the solar cell and the circuit on the system rigidity, and regard the solar wing as a beam of isotropic material with equal section.
(d) The elastic deformation of the solar wing yields assumptions of small deformation and linear conditions, and the materials satisfy Hooke's law.
(e) Considering the eccentricity in the solar wing, the beam section is an asymmetric section. When describing the torsion motion, the warping effect of the section is considered.

(2) **Spinning flexible smart beam**

Furthermore, a piezoelectric structure is pasted on the flexible beam as an active vibration control actuator to simplified to be a spinning flexible smart beam, as shown in Fig. 2.5. The boundary conditions are as described in (1).

The basic assumptions of model simplification are as follows:

(a) The simplified model of flexible beam is based on the assumption in (1).
(b) The mass and stiffness of the adhesive between the beam and PZT patches are neglected.
(c) The strain of the beam's torsional deformation on PZT patches are neglected.

(3) **Beam-plate combined structure model**

For a flexible appendage whose dimension in one direction is much smaller than those in the other two directions (as the solar wing shown in Fig. 2.6a), it is simplified to a spinning combined structure of beam and plate, as shown in Fig. 2.6b. The boundary condition is that the root is fixed and only the degree of freedom of rotation along the unfolding direction is released.

Taking the solar wing as an example, the basic assumptions for model simplification are as follows:

Fig. 2.5 spinning flexible smart beam

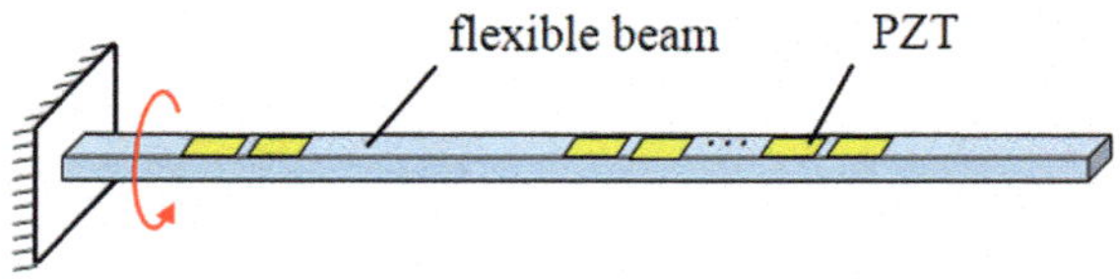

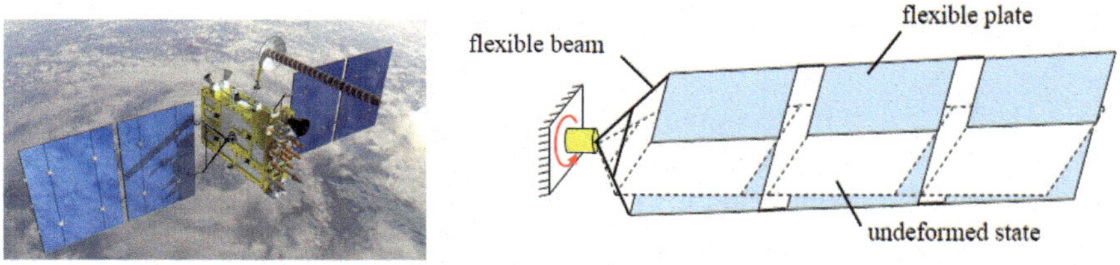

(a) satellite solar panel (b) simplified model

Fig. 2.6 Spinning beam-plate combined structure model

(a) The assumptions are the same as the basic assumptions (a), (b) and (d) in (1).
(b) The yoke is simplified as an isotropic beam, and the solar array is simplified as an isotropic plate, ignoring the influence of solar cells and circuit on the stiffness of the system.
(c) The hinge between the yoke and the solar panel and hinges linked solar panels are rigid connections.

2.2.3.2 Simplified Model of Spacecraft

According to configurations of flexible spacecraft, the spacecraft is simplified to be a system with a central body and one or more appendages, as shown in Fig. 2.7. The basic assumptions for model simplification are as follows:

(1) The central body which represents the spacecraft platform is simplified to be a rigid body, which can perform unrestricted rotations and translations in the inertial space;
(2) The appendage is simplified to be a flexible body, which has only rotations relative to the platform, and degrees of freedom of translations of the hinge point are limited;
(3) Ignore nonlinear factors such as the gap and the friction of the connection between the platform and the appendage;
(4) The elastic deformation yields to small deformation and linear conditions, and physical properties of materials of the system are elastic and constant;
(5) The center of mass of the system remains unchanged.

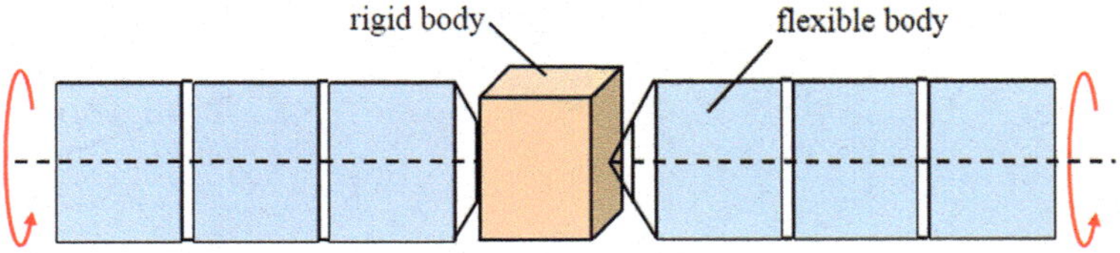

Fig. 2.7 Rigid-flexible coupling model of flexible spacecraft with rotating structures

2.3 Mathematical Model for Flexible Appendages

In this section, based on the simplified model of the spinning flexible appendage in the previous section, a dynamic model of the flexible appendage is established. Both the analytical method and the finite element method are used to establish the dynamic model of the spinning flexible beam. Then the method to calculate the natural frequency and mode shape of the system based on the complex modal theory is given. The piezoelectric sheets are pasted on the beam, and the electromechanical coupling model of the smart flexible beam is established. Finally, to establish the dynamic model for the simplified model of the beam-plate combined structure, the finite element method is utilized to derive the governing equation.

2.3.1 Modeling of Spinning Beam

2.3.1.1 Description of the Spinning Beam

The system to be studied is depicted in Fig. 2.8 in its deformed state. The basic element of the system is a beam that is of length L and is in an asymmetric uniform cross section. When the beam does not deflect, the x-axis is the shear center axis of the beam. The right end of the beam is free, while the left end ($x = 0$) is connected to a base which spins about the longitudinal axis at a constant angular velocity denoted as Ω.

In order to formulate the equations of motion of the spinning beam, a system at reference frames needs to be considered. Three orthogonal right-handed coordinate frames are used in order to describe the position vector $\boldsymbol{R}$ of a differential element dM at a point P.

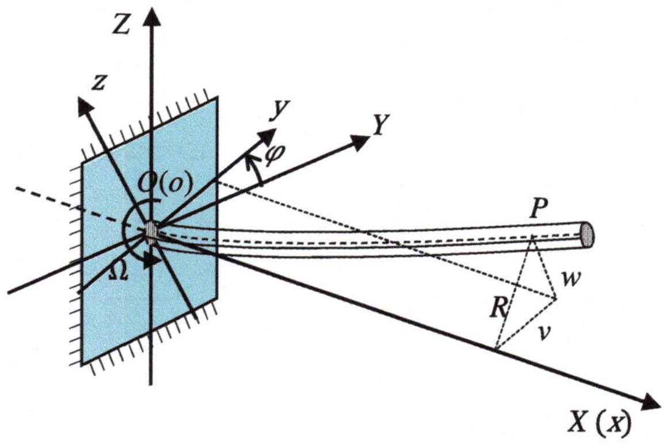

Fig. 2.8 Deformed spinning beam with arbitrary cross section

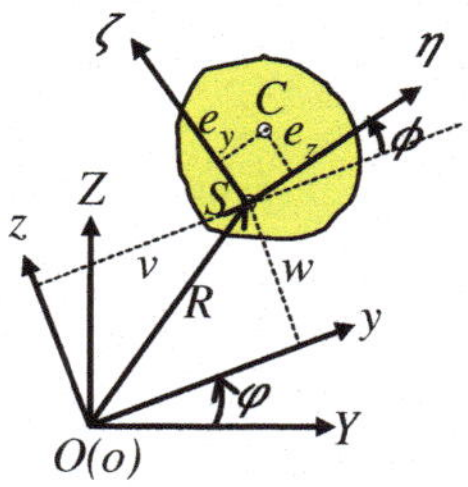

Fig. 2.9 Sketch map of a beam cross section in the deformed state

(a) Frame *OXYZ* is an inertial frame which has an origin fixed at point *O*.
(b) Frame *oxyz* is a body frame, which in its initial state is coincident with the frame of *OXYZ*. During the beam is spinning, the frame *oxyz* is rotated with the x axis in the spinning angle velocity, so that the x axis remains parallel to the X axis of the frame *OXYZ* and the y and z axis are variable. The angle between the y axis and the Y axis is represented by the symbol φ. The origins of the frames *OXYZ* and *oxyz* are coincident, and these origins are fixed to shear center of the beam cross section in the clamped end.
(c) The third reference frame $S\xi\eta\zeta$, is the element coordinate of the differential element which is attached to the shear center S of the beam section. The frame $S\xi\eta\zeta$ is paralleled to the frame *oxyz* when the beam is in its undeformed state. The relative displacement between the frame *oxyz* and the frame $S\xi\eta\zeta$ is equal to the torsion angle at the right end of the beam. C represents the center of mass of the beam section and (e_y, e_z) denotes the coordinate of C in the frame $S\xi\eta\zeta$.

When the flexible beam undergoes a displacement, the section rotates an angle denoted by ϕ. During free vibration, a possible configuration of the system is shown in Fig. 2.9. The deformation of a differential element located at a distance x from the left end is defined by spatial displacement $u(x,t)$, $v(x,t)$, $w(x,t)$ and rotation $\phi(x,t)$ on x-axis. The $u(x,t)$ represents the axial displacement in the x-direction while $v(x,t)$ and $w(x,t)$ represent the lateral displacement in the y-direction and z-direction, respectively.

2.3.1.2 Analytical Model of the Spinning Beam

The beam model is based on the Euler–Bernoulli model, and considers the effect of torsional inertia, and does not consider the effect of shear deformation or axial force on the beam's lateral or torsional deformation.

(1) **Energy equation**

We consider the spinning beam undergoing transverse displacements and torsional motion. Also, it is assumed that the shear center S and the center of mass C of the cross section are not coincident. The position vector of a representative point after beam deformation can be defined as

$$\boldsymbol{R}(x) = u\boldsymbol{i} + v\boldsymbol{j} + w\boldsymbol{k} + e_y\boldsymbol{j}_1 + e_z\boldsymbol{k}_1 \tag{2.1}$$

where $\boldsymbol{i}, \boldsymbol{j}$ and $\boldsymbol{k}$ are unit vectors in the x, y and z directions, respectively. And $\boldsymbol{i_1}, \boldsymbol{j_1}$ and $\boldsymbol{k_1}$ are unit vectors in the ξ, η and ζ directions, respectively.

The velocity of the point can be obtained as follows

$$\boldsymbol{v}(x) = \dot{u}\boldsymbol{i} + \dot{v}\boldsymbol{j} + \dot{w}\boldsymbol{k} + \Omega\boldsymbol{i} \times (v\boldsymbol{j} + w\boldsymbol{k}) + \left(\Omega + \dot{\phi}\right)\boldsymbol{i} \times \left(e_y\boldsymbol{j}_1 + e_z\boldsymbol{k}_1\right) \tag{2.2}$$

$\boldsymbol{v}(x)$ represents the velocity of center. The overhead dot denotes partial derivatives with respect to time t.

The formula for kinetic energy can be simplified as

$$\begin{aligned}
T_b &= \frac{1}{2}\int_0^L \rho A\boldsymbol{v}^2(x)dx \\
&= \frac{1}{2}\int_0^L \rho A\left(\dot{u}^2 + \dot{v}^2 + \dot{w}^2\right)dx + \frac{1}{2}\rho J_p \int_0^L \dot{\phi}^2 dx \\
&\quad + \frac{1}{2}\int_0^L \rho A\left[\Omega^2\left(v^2 + w^2\right) - 2\Omega\dot{v}w + 2\Omega v\dot{w}\right]dx \\
&\quad + \frac{1}{2}\int_0^L \rho A\left[\left(e_y^2 + e_z^2\right)\dot{\phi}^2 - 2e_z\dot{v}\dot{\phi} + 2e_y\dot{w}\dot{\phi} + 2e_z\Omega w\dot{\phi} + 2e_y\Omega v\dot{\phi}\right]dx \\
&\quad + \int_0^L \rho A\left[\left(e_y^2 + e_z^2\right)\Omega\dot{\phi} - e_z\Omega\dot{v} + e_z\Omega^2 w + e_y\Omega\dot{w} + e_y\Omega^2 v\right]dx \\
&\quad + \int_0^L \rho A\left(-e_y\Omega\dot{v}\phi + e_y\Omega^2 w\phi - e_z\Omega\dot{w}\phi - e_z\Omega^2 v\phi\right)dx + \frac{1}{2}\int_0^L \rho Ae^2\Omega^2 dx
\end{aligned} \tag{2.3}$$

The symbols ρ, E and A denote the density, Young's modulus and the cross-sectional area respectively. In Eq. (2.3), the first term on the right side of the equal sign is the kinetic energy generated by the axial stretching vibration and lateral vibration of the beam, the second term is the kinetic energy generated by the axial torsional vibration, and the other terms are the contribution of rotation and eccentricity to the kinetic energy.

The potential strain energy of the beam including the warping effect is considered as below

$$\begin{aligned}
U_b = \frac{1}{2}\int_0^L EAu'^2 dx + \frac{1}{2}\int_0^L E\left(I_z v''^2 + I_y w''^2\right)dx + \frac{1}{2}\int_0^L GJ_p\phi'^2 dx \\
+ \frac{1}{2}\int_0^L E\Gamma\phi''^2 dx
\end{aligned} \tag{2.4}$$

where G denotes the shear modulus. I_y and I_z show the second moments of area about the z-axis and y-axis, $E\Gamma$ is warping rigidity. Primes denote partial derivatives with respect to x. J_p is the polar moment of inertia and is given by

$$J_p = \iint\limits_A r^2 d\eta d\zeta \tag{2.5}$$

where r represents the distance between a certain point in the section and the center.

For uniform beams, A, I_y, I_z, J_p and $E\Gamma$ are constant throughout the span. The first term on the right side of the above equation is the potential energy of axial stretching vibration, the second term is the potential energy of transverse vibration, the third term is the potential energy of axial torsional vibration, and the fourth term is the contribution of warping to potential energy.

(2) **System dynamics equation**

The Lagrange function of the beam can be expressed as

$$L = T_b - U_b \tag{2.6}$$

Using the Hamilton's principle, the dynamic model of the system can be obtained

$$\begin{aligned}
&\frac{E}{\rho}\frac{\partial^2 u}{\partial x^2} - \ddot{u} = 0 \\
&EI_z\frac{\partial^4 v}{\partial x^4} + \rho A\left(\ddot{v} - \Omega^2 v - 2\Omega\dot{w} - e_z\ddot{\phi} - 2\Omega e_y\dot{\phi}e_z\Omega^2\phi\right) = \rho A e_y\Omega^2 \\
&EI_y\frac{\partial^4 w}{\partial x^4} + \rho A\left(\ddot{w} - \Omega^2 w + 2\Omega\dot{v} + e_y\ddot{\phi} - 2\Omega e_z\dot{\phi} - e_y\Omega^2\phi\right) = \rho A e_z\Omega^2 \\
&E\Gamma\frac{\partial^4 \phi}{\partial x^4} - GJ_p\frac{\partial^2 \phi}{\partial x^2} + \left(\rho A e^2 + \rho J_p\right)\ddot{\phi} + \rho A e_y\left(2\Omega\dot{v} + \ddot{w} - \Omega^2 w\right) + \ldots \\
&\qquad \ldots + \rho A e_z\left(-\ddot{v} + \Omega^2 v + 2\Omega\dot{w}\right) = 0
\end{aligned} \tag{2.7}$$

It can be seen from the above formula that the axial stretching vibration of the beam is decoupled from the lateral vibration and the axial torsional vibration, while the lateral vibration and the axial torsional vibration are coupled. Therefore, when calculating the natural frequency and mode shape, the axial stretching vibration can be calculated separately.

When skipping the eccentricity of the cross section, Eq. (2.7) has the following form [19]

$$\frac{E}{\rho}\frac{\partial^2 u}{\partial x^2} - \ddot{u} = 0$$

$$
\begin{aligned}
&EI_z\frac{\partial^4 v}{\partial x^4}+\rho A\left(\ddot{v}-\Omega^2 v-2\Omega\dot{w}\right)=0\\
&EI_y\frac{\partial^4 w}{\partial x^4}+\rho A\left(\ddot{w}-\Omega^2 w+2\Omega\dot{v}\right)=0\\
&E\Gamma\frac{\partial^4\phi}{\partial x^4}-GJ_p\frac{\partial^2\phi}{\partial x^2}+\rho J_p\ddot{\phi}=0
\end{aligned}
\tag{2.8}
$$

In this case, the axial torsional vibration is decoupled from the lateral vibration. It can be seen that the noncoincidence of the shear center and the centroid causes the coupling of the axial torsional vibration and the lateral vibration. Also, it can be concluded that the eccentricity induces the coupling between transverse deformations and torsional motion.

When skipping the spinning, Eq. (2.7) has the following form, which is consistent with the results in Ref. [20]

$$
\begin{aligned}
&\frac{E}{\rho}\frac{\partial^2 u}{\partial x^2}-\ddot{u}=0\\
&EI_z\frac{\partial^4 v}{\partial x^4}+\rho A\left(\ddot{v}-e_z\ddot{\phi}\right)=0\\
&EI_y\frac{\partial^4 w}{\partial x^4}+\rho A\left(\ddot{w}+e_y\ddot{\phi}\right)=0\\
&E\Gamma\frac{\partial^4\phi}{\partial x^4}-GJ_p\frac{\partial^2\phi}{\partial x^2}+\left(\rho Ae^2+\rho J_p\right)\ddot{\phi}+\rho A\left(-e_z\ddot{v}+e_y\ddot{w}\right)=0
\end{aligned}
\tag{2.9}
$$

In this case, the displacement v and w in the two directions of lateral vibration are decoupled. It is obvious that the coupling between v and w takes place due to the spinning.

(3) **Mode shape and frequency equation**

The detailed explanation of the beam's axial stretching vibration can be found in the literature [21]. For a free homogeneous vibration, a sinusoidal oscillation is assumed

$$
\begin{aligned}
v(x,t)&=Y(x)e^{j\omega t}\\
w(x,t)&=Z(x)e^{j\omega t}\\
\phi(x,t)&=\psi(x)e^{j\omega t}
\end{aligned}
\qquad \left(j=\sqrt{-1}\right)
\tag{2.10}
$$

where ω is the circular frequency of oscillation, Y, Z and Ψ are amplitudes of v, w and ϕ, respectively.

Substituting Eq. (2.10) into the differential Eq. (2.7) leads to

$$
\frac{EI_z}{\rho A}Y^{(4)}-\left(\omega^2+\Omega^2\right)Y-2j\omega\Omega Z+e_z\left(\omega^2+\Omega^2\right)\psi-2j\omega\Omega e_y\psi=0
$$

$$\frac{EI_y}{\rho A}Z^{(4)} - \left(\omega^2 + \Omega^2\right)Z + 2j\omega\Omega Y - e_y\left(\omega^2 + \Omega^2\right)\psi - 2j\omega\Omega e_z\psi = 0$$
$$\frac{E\Gamma}{\rho A}\psi^{(4)} - \frac{GJ_P}{\rho A}\psi'' - \left(e^2 + \frac{J_P}{A}\right)\omega^2\psi + \ldots$$
$$\ldots + \left(\omega^2 + \Omega^2\right)\left(e_z Y - e_y Z\right) + 2j\omega\Omega\left(e_y Y + e_z Z\right) = 0 \quad (2.11)$$

For convenience, we consider a beam with a monosymmetric cross-section with symmetric axis y. The centroid C is on the axis y and the scalar e_z is equal to zero. Then Eq. (2.11) can be simplified as

$$\frac{EI_z}{\rho A}Y^{(4)} - (\omega^2 + \Omega^2)Y - 2j\omega\Omega Z - 2j\omega\Omega e_y\psi = 0$$
$$\frac{EI_y}{\rho A}Z^{(4)} - (\omega^2 + \Omega^2)Z + 2j\omega\Omega Y - e_y(\omega^2 + \Omega^2)\psi = 0$$
$$\frac{E\Gamma}{\rho A}\psi^{(4)} - \frac{GJ_P}{\rho A}\psi'' - \left(e^2 + \frac{J_P}{A}\right)\omega^2\psi + 2j\omega\Omega e_y Y - e_y(\omega^2 + \Omega^2)Z = 0$$
$$(2.12)$$

Then introduce the differential operator D and subsequent variables as follows

$$D = d/dx$$
$$L_{11} = \frac{EI_z}{\rho A}D^4 - \left(\omega^2 + \Omega^2\right) \quad L_{12} = -2j\omega\Omega \quad L_{13} = -j\omega\Omega e_y$$
$$L_{21} = 2j\omega\Omega \quad L_{22} = \frac{EI_y}{\rho A}D^4 - \left(\omega^2 + \Omega^2\right) \quad L_{23} = -e_y\left(\omega^2 + \Omega^2\right)$$
$$L_{31} = 2j\omega\Omega e_y \quad L_{32} = -e_y\left(\omega^2 + \Omega^2\right) \quad L_{33} = \frac{E\Gamma}{\rho A}D^4 - \frac{GJ_P}{\rho A}D^2 - \left(e_y^2 + \frac{J_P}{A}\right)\omega^2$$
$$(2.13)$$

Equation (2.12) can be expressed as the following form

$$L_{11}Y + L_{12}Z + L_{13}\psi = 0$$
$$L_{21}Y + L_{22}Z + L_{23}\psi = 0 \quad (2.14)$$
$$L_{31}Y + L_{32}Z + L_{33}\psi = 0$$

It can be seen that Y, Z and Ψ satisfy the equation

$$\Delta\begin{bmatrix} Y \\ Z \\ \psi \end{bmatrix} = 0 \quad (2.15)$$

where

$$\Delta = \begin{bmatrix} L_{11} & L_{12} & L_{13} \\ L_{21} & L_{22} & L_{23} \\ L_{31} & L_{32} & L_{33} \end{bmatrix} \tag{2.16}$$

Introducing

$$\begin{aligned} &\kappa_1 = \frac{\rho A}{EI_z} \qquad \kappa_2 = \frac{\rho A}{EI_y} \qquad \kappa_3 = \frac{\rho A}{E\Gamma} \qquad \kappa_4 = \frac{GJ_p}{E\Gamma} \\ &\kappa_5 = \left(e_y^2 + \frac{J_p}{A}\right) \qquad \kappa_6 = \left(\omega^2 + \Omega^2\right) \end{aligned} \tag{2.17}$$

Setting the determinant of differential operator matrix in Eq. (2.16) equal to zero leads to the following twelve order differential equation

$$\begin{aligned} &\left[D^4 - \kappa_6\kappa_1\right]\left\{\left[D^4 - \kappa_6\kappa_2\right]\left(D^4 - \kappa_4 D^2 - \kappa_3\kappa_5\omega^2\right) - e_y^2\kappa_2\kappa_3\kappa_6^2\right\} \\ &\quad - 4\omega^2\Omega^2\kappa_1\kappa_2\left[\left(D^4 - \kappa_4 D^2 - \kappa_3\kappa_5\omega^2\right) + e_y^2\kappa_3\kappa_6\right] + 4\omega^2\Omega^2 e_y^2\kappa_1\kappa_3 D^4 = 0 \end{aligned} \tag{2.18}$$

The solution of the above equation can be expressed in the exponential form

$$R(\zeta) = e^{rx} \tag{2.19}$$

Specify

$$s = r^2 \tag{2.20}$$

Substituting Eqs. (2.19) and (2.20) into Eq. (2.18), the following characteristic equation can be obtained

$$\begin{aligned} &s^6 - \kappa_4 s^5 - \left(\kappa_3\kappa_5\omega^2 + \kappa_2\kappa_6 + \kappa_1\kappa_6\right)s^4 + (\kappa_2\kappa_4\kappa_6 + \kappa_1\kappa_4\kappa_6)s^3 \\ &+(\kappa_1\kappa_2\kappa_6^2 - 4\kappa_1\kappa_2\omega^2\Omega^2 - \kappa_2\kappa_3\kappa_6^2 e_y^2 - 4\kappa_1\kappa_3 e_y^2\omega^2\Omega^2 + \ldots \\ &+\kappa_1\kappa_3\kappa_5\kappa_6\omega^2 + \kappa_2\kappa_3\kappa_5\kappa_6\omega^2)s^2 + \kappa_1\kappa_2\kappa_4\left(4\omega^2\Omega^2 - \kappa_6^2\right)s \\ &-\kappa_1\kappa_2\kappa_3\left(\kappa_5\kappa_6^2\omega^2 - \kappa_6^3 e_y^2 + 4\kappa_6 e_y^2\omega^2\Omega^2 - 4\kappa_5\omega^4\Omega^2\right) = 0 \end{aligned} \tag{2.21}$$

$s_1 \sim s_6$ are solutions of Eq. (2.21). The twelve roots of Eq. (2.18) can be written as

$$\begin{aligned} &\pm r_1, \quad \pm r_2, \quad \pm r_3, \quad \pm r_4, \quad \pm r_5, \quad \pm r_6 \\ &s.t. \;\; r_1 = j\sqrt[2]{s_1}, \quad r_2 = j\sqrt[2]{s_2}, \quad r_3 = j\sqrt[2]{s_3}, \quad r_4 = \sqrt[2]{s_4}, \quad r_5 = \sqrt[2]{s_5}, \quad r_6 = \sqrt[2]{s_6} \end{aligned} \tag{2.22}$$

Then the general solutions of Y, Z and Ψ are expressed as

$$
\begin{aligned}
Y(\zeta) &= A_1 \cosh r_1 x + A_2 \sinh r_1 x + A_3 \cosh r_2 x + A_4 \sinh r_2 x \\
&\quad + A_5 \cosh r_3 x + A_6 \sinh r_3 x + A_7 \cos r_4 x + A_8 \sin r_4 x \\
&\quad + A_9 \cos r_5 x + A_{10} \sin r_5 x + A_{11} \cos r_6 x + A_{12} \sin r_6 x \\
Z(\zeta) &= B_1 \cosh r_1 x + B_2 \sinh r_1 x + B_3 \cosh r_2 x + B_4 \sinh r_2 x \\
&\quad + B_5 \cosh r_3 x + B_6 \sinh r_3 x + B_7 \cos r_4 x + B_8 \sin r_4 x \\
&\quad + B_9 \cos r_5 x + B_{10} \sin r_5 x + B_{11} \cos r_6 x + B_{12} \sin r_6 x \\
\psi(\zeta) &= C_1 \cosh r_1 x + C_2 \sinh r_1 x + C_3 \cosh r_2 x + C_4 \sinh r_2 x \\
&\quad + C_5 \cosh r_3 x + C_6 \sinh r_3 x + C_7 \cos r_4 x + C_8 \sin r_4 x \\
&\quad + C_9 \cos r_5 x + C_{10} \sin r_5 x + C_{11} \cos r_6 x + C_{12} \sin r_6 x
\end{aligned}
\tag{2.23}
$$

where A_i, B_i and C_i ($i = 1 \sim 12$) are three different sets of constants.

Substituting Eq. (2.23) into Eq. (2.12), relations between A_i, B_i and C_i can be derived

$$
\begin{aligned}
&B_1 = p_1 A_1 \quad B_2 = p_1 A_2 \quad B_3 = p_2 A_3 \quad B_4 = p_2 A_4 \quad B_5 = p_3 A_5 \\
&B_6 = p_3 A_6 \\
&B_7 = p_4 A_7 \quad B_8 = p_4 A_8 \quad B_9 = p_5 A_9 \quad B_{10} = p_5 A_{10} \quad B_{11} = p_6 A_{11} \\
&B_{12} = p_6 A_{12} \\
&C_1 = q_1 A_1 \quad C_2 = q_1 A_2 \quad C_3 = q_2 A_3 \quad C_4 = q_2 A_4 \quad C_5 = q_3 A_5 \\
&C_6 = q_3 A_6 \\
&C_7 = q_4 A_7 \quad C_8 = q_4 A_8 \quad C_9 = q_5 A_9 \quad C_{10} = q_5 A_{10} \quad C_{11} = q_6 A_{11} \\
&C_{12} = q_6 A_{12}
\end{aligned}
\tag{2.24}
$$

where

$$
\begin{aligned}
p_i &= \frac{\kappa_2 \kappa_6}{2 j \omega r_i^4 \Omega} \left(\frac{1}{\kappa_1} r_i^4 - \kappa_6 + \frac{4\omega^2 \Omega^2}{\kappa_6} \right) \qquad (i = 1, 2, \cdots, 6) \\
q_i &= \begin{cases}
\dfrac{\kappa_2 \kappa_6^2 e \left(\frac{1}{\kappa_1} r_i^4 - \kappa_6 + \frac{4\omega^2\Omega^2}{\kappa_6} \right) + 4 e_y \omega^2 \Omega^2 r_i^4}{2 j \omega r_i^4 \Omega \left(\frac{1}{\kappa_3} r_i^4 - \frac{\kappa_4}{\kappa_3} r_i^2 - \kappa_5 \omega^2 \right)} & (i = 1, 2, 3) \\[2ex]
\dfrac{\kappa_2 \kappa_6^2 e \left(\frac{1}{\kappa_1} r_i^4 - \kappa_6 + \frac{4\omega^2\Omega^2}{\kappa_6} \right) + 4 e_y \omega^2 \Omega^2 r_i^4}{2 j \omega r_i^4 \Omega \left(\frac{1}{\kappa_3} r_i^4 + \frac{\kappa_4}{\kappa_3} r_i^2 - \kappa_5 \omega^2 \right)} & (i = 4, 5, 6)
\end{cases}
\end{aligned}
\tag{2.25}
$$

The constants $A_1 \sim A_{12}$, can be determined from the boundary conditions. For a clamped-free beam, the boundaries are as follows

$$\begin{aligned} &clamped\ end \quad (x=0): V=0, V'=0, W=0, W'=0, \Phi=0, \Phi'=0 \\ &free\ end \quad (x=L): V''=0, V'''=0, W''=0, \\ &W'''=0, \kappa_4\Phi'-\Phi'''=0, \Phi''=0 \end{aligned} \tag{2.26}$$

Using the boundary condition, a set of twelve homogeneous equations in terms of the constants $A_1 \sim A_{12}$ will be generated. The natural frequencies ω can be numerically solved by setting the determinant of the coefficient matrix of $A_1 \sim A_{12}$ to equal zero.

2.3.1.3 FEM Model of the Spinning Beam

(1) Discretization of flexible beam.

The flexible beam is divided into multiple elements along the axial direction, as shown in Fig. 2.10. The node displacement vector of the section centroid of the i-th element in the body coordinate system can be expressed as

$$\{\delta\}^e = \left\{u_i\ v_i\ w_i\ \theta_{x,i}\ \theta_{y,i}\ \theta_{z,i}\ \vartheta_i\ u_j\ v_j\ w_j\ \theta_{x,j}\ \theta_{y,j}\ \theta_{z,j}\ \vartheta_j\right\}^T \tag{2.27}$$

Each node has seven degrees of freedom, in which six degrees of freedom are consistent with the degrees of freedom of the classical 3D beam, and the seventh degree of freedom is the warping angle. When the shear deformation of axial torsion is not considered, the warping angle satisfies [22]

$$\vartheta = \theta_x' \tag{2.28}$$

The force acting on the flexible beam can be expressed as

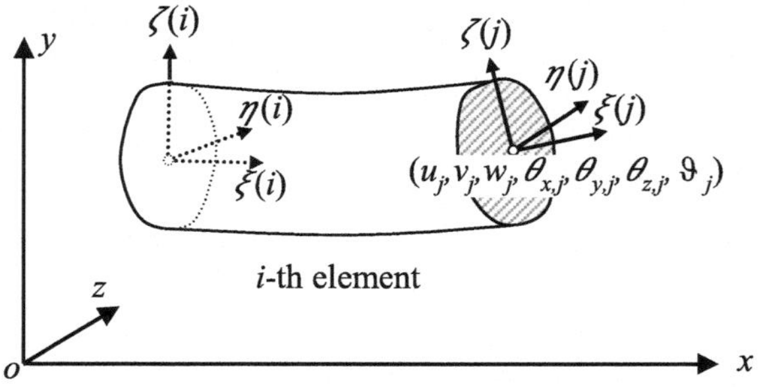

Fig. 2.10 Schematic diagram of flexible beam element

$$\boldsymbol{p}(x,t)=\begin{bmatrix}p_x(x,t)\\ p_y(x,t)\\ p_z(x,t)\\ q_x(x,t)\\ q_y(x,t)\\ q_z(x,t)\end{bmatrix} \tag{2.29}$$

where p_x, p_y, and p_z are the distributed forces acting on the x, y, and z axes, and q_x, q_y, and q_z are the moments acting on the x, y, and z axes, respectively.

The nodal displacement vector can be divided into four parts that are decoupled from each other, namely the axial expansion and contraction displacement, the y-direction bending displacement, the z-direction bending displacement and the axial torsional displacement

$$\begin{aligned}
\{\boldsymbol{\delta}\}_u^e &= \left\{u_i \;\; u_j\right\}^T\\
\{\boldsymbol{\delta}\}_v^e &= \left\{v_i \;\; \theta_{z,i} \;\; v_j \;\; \theta_{z,j}\right\}^T\\
\{\boldsymbol{\delta}\}_w^e &= \left\{w_i \;\; \theta_{y,i} \;\; w_j \;\; \theta_{y,j}\right\}^T\\
\{\boldsymbol{\delta}\}_\vartheta^e &= \left\{\theta_{x,i} \;\; \vartheta_i \;\; \theta_{x,j} \;\; \vartheta_j\right\}^T
\end{aligned} \tag{2.30}$$

The displacement of any point inside the element can be expressed as

$$\begin{array}{lll}
u=\boldsymbol{N}_u\{\boldsymbol{\delta}\}_u^e & v=\boldsymbol{N}_v\{\boldsymbol{\delta}\}_v^e & w=\boldsymbol{N}_w\{\boldsymbol{\delta}\}_w^e\\
\theta_x=\boldsymbol{N}_\vartheta\{\boldsymbol{\delta}\}_\vartheta^e & \theta_y=-\boldsymbol{N'}_w\{\boldsymbol{\delta}\}_w^e & \theta_z=\boldsymbol{N'}_v\{\boldsymbol{\delta}\}_v^e
\end{array} \tag{2.31}$$

In the formula, $\boldsymbol{N}_u$, $\boldsymbol{N}_v$, $\boldsymbol{N}_w$, and $\boldsymbol{N}_\vartheta$ are the shape function matrices of an element, which can be expressed as

$$\begin{aligned}
\boldsymbol{N}_u &= \left[N_1 \; N_2\right]\\
\boldsymbol{N}_v &= \left[N_3 \; -N_4 \; N_5 \; -N_6\right]\\
\boldsymbol{N}_w &= \left[N_3 \; N_4 \; N_5 \; N_6\right]\\
\boldsymbol{N}_\vartheta &= \left[N_3 \; -N_4 \; N_5 \; -N_6\right]
\end{aligned} \tag{2.32}$$

where

$$\begin{array}{ll}
N_1=1-\frac{x}{l} & N_2=\frac{x}{l}\\
N_3=1-\frac{3x^2}{l^2}+\frac{2x^3}{l^3} & N_4=-\left(x-\frac{2x^2}{l}+\frac{x^3}{l^2}\right)\\
N_5=\frac{3x^2}{l^2}-\frac{2x^3}{l^3} & N_6=\frac{x^2}{l}-\frac{x^3}{l^2}
\end{array} \tag{2.33}$$

Therefore, the node displacement vector of the element can be expressed as

$$\begin{bmatrix} u \\ v \\ w \\ \theta_x \\ \theta_y \\ \theta_z \end{bmatrix} = \boldsymbol{N}\{\delta\}^e = \begin{bmatrix} \boldsymbol{N}_u & & & \\ & \boldsymbol{N}_v & & \\ & & \boldsymbol{N}_w & \\ & & & \boldsymbol{N}_\vartheta \\ & & -\boldsymbol{N}'_w & \\ & \boldsymbol{N}'_v & & \end{bmatrix} \begin{bmatrix} \{\delta\}_u^e \\ \{\delta\}_v^e \\ \{\delta\}_w^e \\ \{\delta\}_\vartheta^e \end{bmatrix} \tag{2.34}$$

where $\boldsymbol{N}$ is the shape function matrix of the entire element.

(2) **Energies formulation of element.**

The velocity of the centroid of any section in the element is shown in Eq. (2.2), where θ_x corresponds to ϕ, and the kinetic energy of the i-th element can be expressed as

$$\begin{aligned} T_b^e &= \frac{1}{2}\int_l \rho A \boldsymbol{\upsilon}^T \boldsymbol{\upsilon} dx \\ &= \frac{1}{2}\{\dot{\delta}\}_u^{eT} \boldsymbol{m}_u^e \{\dot{\delta}\}_u^e + \frac{1}{2}\{\dot{\delta}\}_v^{eT} \boldsymbol{m}_v^e \{\dot{\delta}\}_v^e + \frac{1}{2}\{\dot{\delta}\}_w^{eT} \boldsymbol{m}_w^e \{\dot{\delta}\}_w^e \\ &+ \frac{1}{2}\left(1 + \frac{Ae^2}{J_b}\right)\{\dot{\delta}\}_\vartheta^{eT} \boldsymbol{m}_\vartheta^e \{\dot{\delta}\}_\vartheta^e + \frac{1}{2}\Omega^2\{\delta\}_v^{eT} \boldsymbol{m}_v^e \{\delta\}_v^e \\ &+ \frac{1}{2}\Omega^2\{\delta\}_w^{eT} \boldsymbol{m}_w^e \{\delta\}_w^e - \Omega\{\dot{\delta}\}_v^{eT} \boldsymbol{m}_{vw}^e \{\delta\}_w^e + \Omega\{\delta\}_v^{eT} \boldsymbol{m}_{vw}^e \{\dot{\delta}\}_w^e \\ &- e_z\{\dot{\delta}\}_v^{eT} \boldsymbol{m}_{v\vartheta}^e \{\dot{\delta}\}_\vartheta^e + e_y\Omega\{\delta\}_v^{eT} \boldsymbol{m}_{v\vartheta}^e \{\dot{\delta}\}_\vartheta^e - e_y\Omega\{\dot{\delta}\}_v^{eT} \boldsymbol{m}_{v\vartheta}^e \{\delta\}_\vartheta^e \\ &- e_z\Omega^2\{\delta\}_v^{eT} \boldsymbol{m}_{v\vartheta}^e \{\delta\}_\vartheta^e + e_y\{\dot{\delta}\}_w^{eT} \boldsymbol{m}_{w\vartheta}^e \{\dot{\delta}\}_\vartheta^e + e_z\Omega\{\delta\}_w^{eT} \boldsymbol{m}_{w\vartheta}^e \{\dot{\delta}\}_\vartheta^e \\ &- e_z\Omega\{\dot{\delta}\}_w^{eT} \boldsymbol{m}_{w\vartheta}^e \{\delta\}_\vartheta^e + e_y\Omega^2\{\delta\}_w^{eT} \boldsymbol{m}_{w\vartheta}^e \{\delta\}_\vartheta^e + f_v^e\{\delta\}_v^e + f_w^e\{\delta\}_w^e \\ &+ \int_l \rho A\left(e^2\Omega\dot{\theta}_x - e_z\Omega\dot{v} + e_y\Omega\dot{w}\right)dx + \frac{1}{2}\int_l \rho A e^2\Omega^2 dx \end{aligned} \tag{2.35}$$

where

$$\begin{aligned} \boldsymbol{m}_j^e &= \frac{1}{2}\rho A \int_l \boldsymbol{N}_j^T \boldsymbol{N}_j dx \qquad (j = u, v, w) \\ \boldsymbol{m}_j^e &= \frac{1}{2}\rho J_p \int_l \boldsymbol{N}_j^T \boldsymbol{N}_j dx \qquad (j = \vartheta) \\ \boldsymbol{m}_{jk}^e &= \frac{1}{2}\rho A \int_l \boldsymbol{N}_j^T \boldsymbol{N}_k dx \qquad (j, k = v, w, \vartheta \quad k \neq j) \end{aligned} \tag{2.36}$$

and

$$
\begin{aligned}
f_v^e &= \rho A e_y \Omega^2 \int_l \boldsymbol{N}_v dx \\
f_w^e &= \rho A e_y \Omega^2 \int_l \boldsymbol{N}_w dx
\end{aligned} \tag{2.37}
$$

The potential energy of this element can be expressed as

$$
\begin{aligned}
U_b^e &= \frac{1}{2}\int_l EAu'^2 dx + \frac{1}{2}\int_l E\left(I_z v''^2 + I_y w''^2\right) dx \\
&\quad + \frac{1}{2}\int_l GJ_b\theta_x'^2 dx + \frac{1}{2}\int_l E\Gamma\theta_x''^2 dx \\
&= \frac{1}{2}\{\boldsymbol{\delta}\}_u^{eT}\boldsymbol{k}_u^e\{\boldsymbol{\delta}\}_u^e + \frac{1}{2}\{\boldsymbol{\delta}\}_v^{eT}\boldsymbol{k}_v^e\{\boldsymbol{\delta}\}_v^e \\
&\quad + \frac{1}{2}\{\boldsymbol{\delta}\}_w^{eT}\boldsymbol{k}_w^e\{\boldsymbol{\delta}\}_w^e + \frac{1}{2}\{\boldsymbol{\delta}\}_\vartheta^{eT}\left(\boldsymbol{k}_\vartheta^e + \boldsymbol{k}_\vartheta^e\right)\{\boldsymbol{\delta}\}_\vartheta^e
\end{aligned} \tag{2.38}
$$

where

$$
\begin{aligned}
&\boldsymbol{k}_u^e = EA\int_l \boldsymbol{N}'^T_u \boldsymbol{N}'_u dx \qquad \boldsymbol{k}_v^e = EI_z\int_l \boldsymbol{N}''^T_v \boldsymbol{N}''_v dx \\
&\boldsymbol{k}_w^e = EI_y\int_l \boldsymbol{N}''^T_w \boldsymbol{N}''_w dx \qquad \boldsymbol{k}_\vartheta^e = GJ_b\int_l \boldsymbol{N}'^T_\vartheta \boldsymbol{N}'_\vartheta dx \\
&\boldsymbol{k}_{\vartheta\vartheta}^e = E\Gamma\int_l \boldsymbol{N}''^T_\vartheta \boldsymbol{N}''_\vartheta dx
\end{aligned} \tag{2.39}
$$

The external virtual work acting on the element is

$$
W_b^e = \int_l \boldsymbol{p}^T(x,t)\left[u\ v\ w\ \theta_x\ \theta_y\ \theta_z\right]^T dx = \{\boldsymbol{\delta}\}^{eT}\boldsymbol{p}^e \tag{2.40}
$$

where

$$
\boldsymbol{p}^e = \int_l \boldsymbol{N}^T \boldsymbol{p}(x,t) dx \tag{2.41}
$$

(3) **System dynamics equation**

According to Hamilton principle

$$
\delta\int_{t_0}^{t_1}\left(T_b^e - U_b^e + W_b^e\right)dt = 0 \tag{2.42}
$$

In the formula, δ is the variational sign, and the balance equation of the element can be obtained

$$\boldsymbol{m}^e\left\{\ddot{\boldsymbol{\delta}}\right\}^e + \boldsymbol{c}^e\left\{\dot{\boldsymbol{\delta}}\right\}^e + \boldsymbol{k}^e\{\boldsymbol{\delta}\}^e + \boldsymbol{f}^e = \boldsymbol{p}^e \tag{2.43}$$

The element mass matrix is

$$\boldsymbol{m}^e = \begin{bmatrix} \boldsymbol{m}_u^e & & & \\ & \boldsymbol{m}_v^e & & -e_z\boldsymbol{m}_{v\vartheta}^e \\ & & \boldsymbol{m}_w^e & e_y\boldsymbol{m}_{w\vartheta}^e \\ & -e_z\boldsymbol{m}_{\vartheta v}^e & e_y\boldsymbol{m}_{\vartheta w}^e & \left(1+\frac{Ae^2}{J_b}\right)\boldsymbol{m}_\vartheta^e \end{bmatrix} \tag{2.44}$$

The element gyro matrix is

$$\boldsymbol{c}^e = \begin{bmatrix} 0 & & & \\ & & -2\Omega\boldsymbol{m}_{vw}^e & -2e_y\Omega\boldsymbol{m}_{v\vartheta}^e \\ & 2\Omega\boldsymbol{m}_{wv}^e & & -2e_z\Omega\boldsymbol{m}_{w\vartheta}^e \\ & 2e_y\Omega\boldsymbol{m}_{\vartheta v}^e & 2e_z\Omega\boldsymbol{m}_{\vartheta w}^e & \end{bmatrix} \tag{2.45}$$

The element stiffness matrix is

$$\boldsymbol{k}^e = \begin{bmatrix} \boldsymbol{k}_u^e & & & \\ & \boldsymbol{k}_v^e - \Omega^2\boldsymbol{m}_v^e & & e_z\Omega^2\boldsymbol{m}_{v\vartheta}^e \\ & & \boldsymbol{k}_w^e - \Omega^2\boldsymbol{m}_w^e & -e_y\Omega^2\boldsymbol{m}_{w\vartheta}^e \\ & e_z\Omega^2\boldsymbol{m}_{\vartheta v}^e & -e_y\Omega^2\boldsymbol{m}_{\vartheta w}^e & \boldsymbol{k}_\vartheta^e + \boldsymbol{k}_{\vartheta\vartheta}^e \end{bmatrix} \tag{2.46}$$

The gyro force is

$$\boldsymbol{f}^e = \left[0 \; -f_v^e \; -f_w^e \; 0\right]^T \tag{2.47}$$

Combine the dynamic balance equations of all elements to obtain the dynamic equation of the system

$$M\left\{\ddot{\delta}\right\} + \boldsymbol{C}\left\{\dot{\delta}\right\} + \boldsymbol{K}\{\delta\} + \boldsymbol{F} = \boldsymbol{P} \tag{2.48}$$

where

$$\boldsymbol{M} = \sum_{e=1} \boldsymbol{m}^e \quad \boldsymbol{C} = \sum_{e=1} \boldsymbol{c}^e \quad \boldsymbol{K} = \sum_{e=1} \boldsymbol{k}^e \quad \boldsymbol{F} = \sum_{e=1} \boldsymbol{f}^e \quad \boldsymbol{P} = \sum_{e=1} \boldsymbol{p}^e \tag{2.49}$$

Assuming a total of N degrees of freedom in the flexible beam, dimensions of the matrices $\boldsymbol{M}$, $\boldsymbol{C}$ and $\boldsymbol{K}$ in the above formula are all $N \times N$, and the dimensions of the vectors $\boldsymbol{\delta}$, $\boldsymbol{F}$ and $\boldsymbol{P}$ are all $N \times 1$. The matrices $\boldsymbol{M}$, $\boldsymbol{C}$, and $\boldsymbol{K}$ are non-diagonal, so the lateral vibration and axial torsional vibration in the y and z directions are coupled with each other.

(4) Natural frequency and complex mode

The system equations are casted into the state-space format as follows

$$\boldsymbol{A}_{\delta}\dot{\boldsymbol{x}}_{\delta} + \boldsymbol{B}_{\delta}\boldsymbol{x}_{\delta} = \boldsymbol{P}_{\delta} + \boldsymbol{Q}_{\delta} \tag{2.50}$$

where

$$\begin{gathered}\boldsymbol{x}_{\delta} = \begin{bmatrix}\{\dot{\delta}\}\\ \{\delta\}\end{bmatrix} \quad \boldsymbol{A}_{\delta} = \begin{bmatrix}\boldsymbol{0} & \boldsymbol{M}\\ \boldsymbol{M} & \boldsymbol{C}\end{bmatrix} \quad \boldsymbol{B}_{\delta} = \begin{bmatrix}-\boldsymbol{M} & \boldsymbol{0}\\ \boldsymbol{0} & \boldsymbol{K}\end{bmatrix}\\ \boldsymbol{P}_{\delta} = \begin{bmatrix}\boldsymbol{0}\\ \boldsymbol{P}\end{bmatrix} \quad \boldsymbol{Q}_{\delta} = \begin{bmatrix}\boldsymbol{0}\\ -\boldsymbol{f}\end{bmatrix}\end{gathered} \tag{2.51}$$

The adjoint system of Eq. (2.50) is

$$\boldsymbol{A}_{\delta}^{T}\dot{\boldsymbol{x}}_{\delta}^{T} + \boldsymbol{B}_{\delta}^{T}\boldsymbol{x}_{\delta}^{T} = \boldsymbol{P}_{\delta}^{T} + \boldsymbol{Q}_{\delta}^{T} \tag{2.52}$$

The superscript T represents the transformation rank, and H represents the conjugate transformation rank.

The free vibration equation of the above system and its conjugate system can be expressed as

$$\boldsymbol{A}_{\delta}\dot{\boldsymbol{x}}_{\delta} + \boldsymbol{B}_{\delta}\boldsymbol{x}_{\delta} = 0 \qquad \boldsymbol{A}_{\delta}^{T}\dot{\boldsymbol{x}}_{\delta}^{*} + \boldsymbol{B}_{\delta}^{T}\boldsymbol{x}_{\delta}^{*} = 0 \tag{2.53}$$

The asterisk indicates the conjugate of a scalar, vector, or matrix. The solution of the above two systems can be expressed as

$$\boldsymbol{x}_{\delta} = \boldsymbol{\Phi}e^{\lambda t} \qquad \boldsymbol{x}_{\delta}^{*} = \boldsymbol{\psi}e^{\lambda t} \tag{2.54}$$

Substituting the above formula into Eq. (2.53), we can get

$$\boldsymbol{\lambda}\boldsymbol{A}_{\delta}\boldsymbol{\Phi} + \boldsymbol{B}_{\delta}\boldsymbol{\Phi} = 0 \qquad \boldsymbol{\lambda}\boldsymbol{A}_{\delta}^{T}\boldsymbol{\psi} + \boldsymbol{B}_{\delta}^{T}\boldsymbol{\psi} = 0 \tag{2.55}$$

The matrix $\boldsymbol{A}_{\delta}$ is not symmetric, so the eigenvalues $\boldsymbol{\lambda}$, the left eigenvectors $\boldsymbol{\Phi}$ and right eigenvectors $\boldsymbol{\Psi}$ are all complex and conjugate pairs and can be expressed as [23]

$$\begin{aligned}\boldsymbol{\lambda} &= diag(\lambda_1, \lambda_2, \cdots, \lambda_N, \lambda_1^{*}, \lambda_2^{*}, \cdots, \lambda_N^{*})\\ \boldsymbol{\Phi} &= [\phi_1, \phi_2, \cdots, \phi_N, \phi_1^{*}, \phi_2^{*}, \cdots, \phi_N^{*}]\\ \boldsymbol{\psi} &= [\psi_1, \psi_2, \cdots, \psi_N, \psi_1^{*}, \cdots, \psi_2^{*}, \psi_N^{*}]\end{aligned} \tag{2.56}$$

where

$$\begin{aligned}
&\lambda_i = \lambda_i^R \pm j\lambda_i^I (i = 1, 2, \cdots, N) \\
&\phi_i = \phi_i^R \pm j\phi_i^I (i = 1, 2, \cdots, N) \qquad \left(j = \sqrt{-1}\right) \\
&\psi_i = \psi_i^R \pm j\psi_i^I (i = 1, 2, \cdots, N)
\end{aligned} \tag{2.57}$$

Those matrices may be partitioned as

$$\boldsymbol{\lambda} = \begin{bmatrix} \boldsymbol{\lambda}_{1\sim N} & \\ & \boldsymbol{\lambda}_{1\sim N}^* \end{bmatrix} \qquad \boldsymbol{\Phi} = \begin{bmatrix} \boldsymbol{\Phi}_{1\sim N} & \boldsymbol{\Phi}_{1\sim N}^* \end{bmatrix} \qquad \boldsymbol{\psi} = \begin{bmatrix} \boldsymbol{\psi}_{1\sim N} & \boldsymbol{\psi}_{1\sim N}^* \end{bmatrix} \tag{2.58}$$

The left eigenvectors and the right eigenvalues satisfy the orthogonal conditions

$$\begin{aligned}
&\psi_r^T \boldsymbol{A}_\delta \phi_s = \begin{cases} 0(r \neq s) \\ a_r(r = s) \end{cases} \quad \psi_r^H \boldsymbol{A}_\delta \phi_s^* = \begin{cases} 0(r \neq s) \\ a_r^*(r = s) \end{cases} \\
&\psi_r^T \boldsymbol{B}_\delta \phi_s = \begin{cases} 0(r \neq s) \\ b_r(r = s) \end{cases} \quad \psi_r^H \boldsymbol{B}_\delta \phi_s^* = \begin{cases} 0(r \neq s) \\ b_r^*(r = s) \end{cases}
\end{aligned} \tag{2.59}$$

yielding

$$\lambda_r = -b_r/a_r \qquad \lambda_r^* = -b_r^*/a_r^* \tag{2.60}$$

Define $\boldsymbol{\Phi}$ as the modal matrix and define the transformation

$$\boldsymbol{x}_\delta = \boldsymbol{\Phi}\boldsymbol{x}_q = \begin{bmatrix} \boldsymbol{\Phi}_{1\sim N} & \boldsymbol{\Phi}_{1\sim N}^* \end{bmatrix} \begin{bmatrix} \boldsymbol{x}_{1\sim N} \\ \boldsymbol{x}_{1\sim N}^* \end{bmatrix} \tag{2.61}$$

The elements of the vector $\boldsymbol{x}$ all always real, so the modal coordinate z can be divided into two conjugate sub-vectors. Substituting Eq. (2.61) into Eq. (2.50) and considering the conditions of Eq. (2.59), the system equation is

$$\dot{\boldsymbol{x}}_q = \boldsymbol{\lambda}\boldsymbol{x}_q + \boldsymbol{\Pi}\left(\boldsymbol{P}_\delta + \boldsymbol{Q}_\delta\right) \tag{2.62}$$

where

$$\boldsymbol{\Pi} = \begin{bmatrix} \boldsymbol{\Pi}_{1\sim N} \\ \boldsymbol{\Pi}_{1\sim N}^* \end{bmatrix} = \left[\psi_1/a_1, \psi_2/a_2, \cdots, \psi_N/a_N, \psi_1^*/a_1^*, \psi_2^*/a_2^*, \cdots, \psi_N^*/a_N^*\right]^T \tag{2.63}$$

Since the lower modes contribute much to the system, only the first $2n$ modes $\boldsymbol{\Phi}_c$ is retained. Then the system can be written as

$$\dot{\boldsymbol{x}} = \boldsymbol{A}\boldsymbol{x} + \boldsymbol{\Pi}_c\left(\boldsymbol{P}_\delta + \boldsymbol{Q}_\delta\right) \tag{2.64}$$

where

$$\boldsymbol{x} = \begin{bmatrix} \boldsymbol{x}_{1\sim n} \\ \boldsymbol{x}^*_{1\sim n} \end{bmatrix} \quad \boldsymbol{A} = \boldsymbol{\lambda}_c = \begin{bmatrix} \boldsymbol{\lambda}_{1\sim n} & \\ & \boldsymbol{\lambda}^*_{1\sim n} \end{bmatrix} \quad \boldsymbol{\Pi}_c = \begin{bmatrix} \boldsymbol{\Pi}_{1\sim n} \\ \boldsymbol{\Pi}^*_{1\sim n} \end{bmatrix} \tag{2.65}$$

$\boldsymbol{\lambda}_c$ is the reduced eigenvalue matrix, and the reduced mode matrix is

$$\boldsymbol{\Phi}_c = \begin{bmatrix} \boldsymbol{\Phi}_{1\sim n} & \boldsymbol{\Phi}^*_{1\sim n} \end{bmatrix} \qquad \boldsymbol{\psi}_c = \begin{bmatrix} \boldsymbol{\psi}_{1\sim n} & \boldsymbol{\psi}^*_{1\sim n} \end{bmatrix} \tag{2.66}$$

2.3.2 *Modeling of Spinning Smart Beam*

The system studied in this section is shown in Fig. 2.11. Several pairs of piezoelectric patches distributed along the longitudinal axis are bonded on the top and bottom surfaces of the spinning beam. The piezoelectric patch is a single piezoelectric layer of uniform thickness h_p. The piezoelectric patch is polarized along the h_p direction and the piezoelectric material is subject to an electric field E_3 parallel to the polarization. Some patches are directly bonded to the beam surface and acted as sensors, providing sensing by the sensor voltage V_s. The other patches are acted as actuators, which as an active constraining layer that is activated by the control voltage V_a. With appropriate strain control, through proper manipulation of V_s, the piezoelectric layers actuated by controlled voltages generate deformation so that the energy dissipation mechanism can be enhanced and the structural vibration can be damped out.

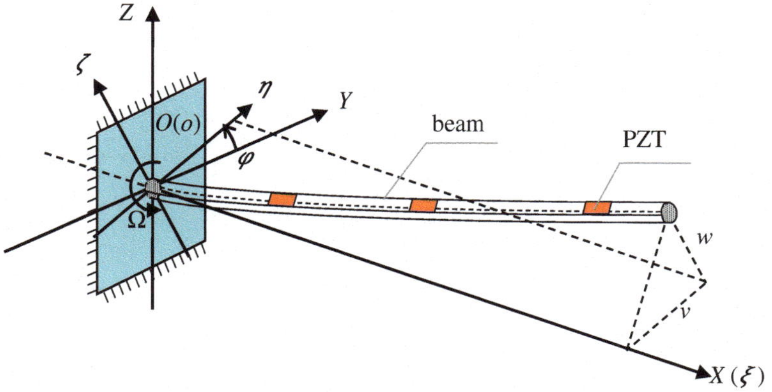

Fig. 2.11 Deformed spinning smart beam with PZT patches

2.3.2.1 Energies of the Spinning Beam with PZT Patches

The flexible beam is first divided into multiple elements. The kinetic energy, potential energy and virtual work of the i-th element containing the piezoelectric sheet can be expressed as

$$\begin{aligned} T^e &= T_b^e + T_p^e \\ U^e &= U_b^e + U_p^e \\ W^e &= W_b^e + W_p^e \end{aligned} \tag{2.67}$$

where ${T_b}^e$ and ${T_p}^e$ represent the kinetic energy of the flexible beam and the piezoelectric sheet respectively, ${U_b}^e$ and ${U_p}^e$ represent the potential energy of the flexible beam and the piezoelectric sheet respectively, ${W_b}^e$ is the external virtual work except the piezoelectric sheet, and ${W_p}^e$ is the external virtual work by piezoelectric sheet. The kinetic energy ${T_b}^e$, potential energy ${U_b}^e$ and external virtual work ${W_b}^e$ of the flexible beam are shown in Eqs. (2.35), (2.38) and (2.40) respectively.

Ignoring the warping of the piezoelectric sheet, the kinetic energy of the piezoelectric sheet on the i-th element can be expressed as

$$\begin{aligned} T^e &= T_b^e + T_p^e \\ &= \frac{1}{2}\int_l \rho A \boldsymbol{\upsilon}^T \boldsymbol{\upsilon} dx + \frac{1}{2}\int_{l_i} \rho_p A_p \boldsymbol{\upsilon}^T \boldsymbol{\upsilon} dx \\ &= \frac{1}{2}\{\dot{\delta}\}_u^{eT} \boldsymbol{m}_u^e \{\dot{\delta}\}_u^e + \frac{1}{2}\{\dot{\delta}\}_v^{eT} \boldsymbol{m}_v^e \{\dot{\delta}\}_v^e + \frac{1}{2}\{\dot{\delta}\}_w^{eT} \boldsymbol{m}_w^e \{\dot{\delta}\}_w^e + \frac{1}{2}\left(1 + \frac{Ae^2}{J_b}\right)\{\dot{\delta}\}_\vartheta^{eT} \boldsymbol{m}_\vartheta^e \{\dot{\delta}\}_\vartheta^e \\ &\quad + \frac{1}{2}\Omega^2\{\delta\}_v^{eT} \boldsymbol{m}_v^e \{\delta\}_v^e + \frac{1}{2}\Omega^2\{\delta\}_w^{eT} \boldsymbol{m}_w^e \{\delta\}_w^e - \Omega\{\dot{\delta}\}_v^{eT} \boldsymbol{m}_{vw}^e \{\delta\}_w^e + \Omega\{\delta\}_v^{eT} \boldsymbol{m}_{vw}^e \{\dot{\delta}\}_w^e \\ &\quad - e_z\{\dot{\delta}\}_v^{eT} \boldsymbol{m}_{v\vartheta}^e \{\dot{\delta}\}_\vartheta^e + e_y\Omega\{\delta\}_v^{eT} \boldsymbol{m}_{v\vartheta}^e \{\dot{\delta}\}_\vartheta^e - e_y\Omega\{\dot{\delta}\}_v^{eT} \boldsymbol{m}_{v\vartheta}^e \{\delta\}_\vartheta^e - e_z\Omega^2\{\delta\}_v^{eT} \boldsymbol{m}_{v\vartheta}^e \{\delta\}_\vartheta^e \\ &\quad + e_y\{\dot{\delta}\}_w^{eT} \boldsymbol{m}_{w\vartheta}^e \{\dot{\delta}\}_\vartheta^e + e_z\Omega\{\delta\}_w^{eT} \boldsymbol{m}_{w\vartheta}^e \{\dot{\delta}\}_\vartheta^e - e_z\Omega\{\dot{\delta}\}_w^{eT} \boldsymbol{m}_{w\vartheta}^e \{\delta\}_\vartheta^e + e_y\Omega^2\{\delta\}_w^{eT} \boldsymbol{m}_{w\vartheta}^e \{\delta\}_\vartheta^e \\ &\quad + f_v^e\{\delta\}_v^e + f_w^e\{\delta\}_w^e + \int_l \rho A\left(e^2\Omega\dot{\theta}_x - e_z\Omega\dot{v} + e_y\Omega\dot{w}\right)dx + \frac{1}{2}\int_l \rho A e^2\Omega^2 dx \\ &\quad + \int_{l_i} \rho_p A_p\left(e^2\Omega\dot{\theta}_x - e_z\Omega\dot{v} + e_y\Omega\dot{w}\right)dx + \frac{1}{2}\int_{l_i} \rho_p A_p e^2\Omega^2 dx \end{aligned} \tag{2.68}$$

where

$$\boldsymbol{m}_j^e = \frac{1}{2}\rho A\int_l \boldsymbol{N}_j^T \boldsymbol{N}_j dx + \frac{1}{2}\rho_p A_p \int_{l_p} \boldsymbol{N}_j^T \boldsymbol{N}_j dx \qquad (j = u, v, w)$$

$$\boldsymbol{m}_j^e = \frac{1}{2}\rho J_p \int_l \boldsymbol{N}_j^T \boldsymbol{N}_j dx \qquad (j = \vartheta)$$

$$\boldsymbol{m}_{jk}^{e} = \begin{cases} \frac{1}{2}\rho A \int_{l} \boldsymbol{N}_j^T \boldsymbol{N}_k dx + \frac{1}{2}\rho_p A_p \int_{l_p} \boldsymbol{N}_j^T \boldsymbol{N}_k dx & (j,k = v,w \quad k \neq j) \\ \frac{1}{2}\rho A \int_{l} \boldsymbol{N}_j^T \boldsymbol{N}_k dx & (j,k = v,w,\vartheta \quad k \neq j) - (j,k = v,w \quad k \neq j) \end{cases} \tag{2.69}$$

and

$$\begin{aligned} f_v^e &= \rho A e_y \Omega^2 \int_{l} \boldsymbol{N}_v dx + \rho_p A_p e_y \Omega^2 \int_{l_p} \boldsymbol{N}_v dx \\ f_w^e &= \rho A e_y \Omega^2 \int_{l} \boldsymbol{N}_w dx + \rho_p A_p e_y \Omega^2 \int_{l_p} \boldsymbol{N}_w dx \end{aligned} \tag{2.70}$$

where l_p and A_p are the length and cross sectional area of the PZT patches, respectively. ρ_p is the material density of the PZT patch.

In the process to formulate governing equations of the beam, we assume the PZT patches are pasted in the z-direction face of the beam. According to the Euler–Bernoulli assumption, the stress and strain fields are uniaxial, along Ox; the axial strain S_1 is related to the curvature w'' by

$$S_1 = -z w'' \tag{2.71}$$

where z is the distance to the neutral axis.

The potential strain energy of the sensor/actuator pairs is considered as below

$$\begin{aligned} U_p^e &= \frac{1}{2}\int_{l_p} dx \int_{A_p} E_p z^2 w''^2 dA \\ &= \frac{1}{2} E_p A_p \left(z_p^2 + z_p t_p + \frac{t_p^2}{3} \right) \int_{l_p} w''^2 dx \end{aligned} \tag{2.72}$$

where t_p is the thickness of each PZT element, z_p is the distance measured from the neutral axis of the beam to the mid-plane of the PZT layer, and E_p is the Young's modulus of the PZT patch.

The potential strain energy of the flexible structure including sensor/actuator pairs is considered as below

$$\begin{aligned} U^e &= U_b^e + U_p^e \\ &= \frac{1}{2}\int_{l} EAu'^2 dx + \frac{1}{2}\int_{l} E\left(I_z v''^2 + I_y w''^2\right) dx + \frac{1}{2}\int_{l} G J_b \theta_x'^2 dx \end{aligned}$$

$$
\begin{aligned}
&+\frac{1}{2}\int_l E\Gamma\theta_x''^2 dx+\frac{1}{2}E_pA_p\left(z_p^2+z_pt_p+\frac{t_p^2}{3}\right)\int_{lp}w''^2dx\\
&=\frac{1}{2}\{\boldsymbol{\delta}\}_u^{eT}\boldsymbol{k}_u^e\{\boldsymbol{\delta}\}_u^e+\frac{1}{2}\{\boldsymbol{\delta}\}_v^{eT}\boldsymbol{k}_v^e\{\boldsymbol{\delta}\}_v^e+\frac{1}{2}\{\boldsymbol{\delta}\}_w^{eT}\boldsymbol{k}_w^e\{\boldsymbol{\delta}\}_w^e\\
&+\frac{1}{2}\{\boldsymbol{\delta}\}_\vartheta^{eT}\left(\boldsymbol{k}_\vartheta^e+\boldsymbol{k}_{\vartheta\vartheta}^e\right)\{\boldsymbol{\delta}\}_\vartheta^e+\frac{1}{2}\{\boldsymbol{\delta}\}_w^{eT}\boldsymbol{k}_p^e\{\boldsymbol{\delta}\}_w^e
\end{aligned}
\tag{2.73}
$$

where

$$
\begin{aligned}
&\boldsymbol{k}_u^e=EA\int_l \boldsymbol{N}'^T_u\boldsymbol{N}'_u dx && \boldsymbol{k}_v^e=EI_z\int_l \boldsymbol{N}''^T_v\boldsymbol{N}''_v dx\\
&\boldsymbol{k}_w^e=EI_y\int_l \boldsymbol{N}''^T_w\boldsymbol{N}''_w dx && \boldsymbol{k}_\vartheta^e=GJ_b\int_l \boldsymbol{N}'^T_\vartheta\boldsymbol{N}'_\vartheta dx\\
&\boldsymbol{k}_\vartheta^e=E\Gamma\int_l \boldsymbol{N}''^T_\vartheta\boldsymbol{N}''_\vartheta dx && \boldsymbol{k}_p^e=E_pA_p\left(z_p^2+z_pt_p+\frac{t_p^2}{3}\right)\int_{l_p}\boldsymbol{N}''^T_w\boldsymbol{N}''_w dx
\end{aligned}
\tag{2.74}
$$

We also assume that the piezoelectric layer is thin enough, so that the applied electrical field density E_3 is constant across the thickness. The virtual work by the PZT patches is the combination of the conservative and non-conservative work terms, defined by integration over the volume of the PZT patches such that

$$
\begin{aligned}
W_p^e&=\frac{1}{2}\int_{l_i}dx\int_{A_p}\left(\varepsilon_{33}E_3^2-2w''ze_{31}E_3\right)dA\\
&=\frac{1}{2}A_pl_i\varepsilon_{33}E_3^2-A_pe_{31}E_3\left(z_p+\frac{t_p}{2}\right)\int_{l_i}w''dx\\
&=\frac{1}{2}\frac{A_pl_i\varepsilon_{33}}{t_p^2}V_i^2-\frac{A_pe_{31}}{t_p}\left(z_p+\frac{t_p}{2}\right)V_i\int_{l_i}\boldsymbol{N}''_w dx\{\boldsymbol{\delta}\}_w^e
\end{aligned}
\tag{2.75}
$$

where ε_{33} is the permittivity, V_i is the electrode voltage acting on the piezoelectric sheet of the element, and the relationship with the electric field density can be expressed as

$$
V_i=t_pE_3 \tag{2.76}
$$

The virtual work of the external force acting on the element is

$$
\begin{aligned}
W^e&=W_b^e+W_p^e\\
&=\{\boldsymbol{\delta}\}^{eT}\boldsymbol{p}^e+\frac{1}{2}\frac{A_pl_i\varepsilon_{33}}{t_p^2}V_i^2-\frac{A_pe_{31}}{t_p}\left(z_p+\frac{t_p}{2}\right)V_i\int_{l_i}\boldsymbol{N}''_w dx\{\boldsymbol{\delta}\}_w^e
\end{aligned}
$$

$$= \{\delta\}^{eT} \boldsymbol{p}^e + \frac{1}{2} V_i^T \gamma_i^e V_i - V_i \boldsymbol{\chi}^e \{\delta\}^e \tag{2.77}$$

with

$$\gamma_i = \frac{l_i A_p}{t_p^2} \varepsilon_{33}$$
$$\boldsymbol{\chi}^e = \left[0 \; 0 \; \frac{A_p e_{31}}{t_p} \left(z_p + \frac{t_p}{2} \right) \int_{l_i} \boldsymbol{N}''_w dx \; 0 \right] \tag{2.78}$$

2.3.2.2 Electromechanical Coupling Equation

Using the Hamilton's principle, the dynamic model of the element can be obtained in the following form

$$\begin{gathered} \boldsymbol{m}^e \left\{ \ddot{\delta} \right\}^e + \boldsymbol{c}^e \left\{ \dot{\delta} \right\}^e + \boldsymbol{k}^e \{\delta\}^e + \boldsymbol{f}^e = \boldsymbol{p}^e + \boldsymbol{\chi}^{eT} V^e \\ \gamma_i V_i = \boldsymbol{\chi}^e \{\delta\}^e \end{gathered} \tag{2.79}$$

The mass matrix and the gyro matrix of the element are consistent with Eqs. (2.44) and (2.45). The stiffness matrix is

$$\boldsymbol{k}^e = \begin{bmatrix} \boldsymbol{k}_u^e & & & \\ & \boldsymbol{k}_v^e - \Omega^2 \boldsymbol{m}_v^e & & e_z \Omega^2 \boldsymbol{m}_{v\vartheta}^e \\ & & \boldsymbol{k}_w^e - \Omega^2 \boldsymbol{m}_w^e + \boldsymbol{k}_p^e & -e_y \Omega^2 \boldsymbol{m}_{w\vartheta}^e \\ & e_z \Omega^2 \boldsymbol{m}_{\vartheta v}^e & -e_y \Omega^2 \boldsymbol{m}_{\vartheta w}^e & \boldsymbol{k}_\vartheta^e + \boldsymbol{k}_{\vartheta\vartheta}^e \end{bmatrix} \tag{2.80}$$

The governing equations of the whole structure by assembling the element matrix into global matrix directly can be obtained as follows

$$\begin{gathered} \boldsymbol{M} \left\{ \ddot{\delta} \right\} + \boldsymbol{C} \left\{ \dot{\delta} \right\} + \boldsymbol{K} \{\delta\} + \boldsymbol{F} = \boldsymbol{P} + \boldsymbol{\chi}^T \boldsymbol{V} \\ \boldsymbol{V} = \boldsymbol{\gamma}^{-1} \boldsymbol{\chi} \{\delta\} \end{gathered} \tag{2.81}$$

with

$$\boldsymbol{V} = [V_1 \; V_2 \cdots]^T \qquad \boldsymbol{\gamma} = diag(\gamma_i) \qquad \boldsymbol{\chi} = \sum_{e=1}^{n} \boldsymbol{\chi}^e \tag{2.82}$$

In a flexible beam, some piezoelectric sheets act as actuators and the others act as sensors. Therefore, the governing equation can be expressed as

$$\begin{aligned} \boldsymbol{M}\left\{\ddot{\delta}\right\} + \boldsymbol{C}\left\{\dot{\delta}\right\} + \boldsymbol{K}\{\delta\} + \boldsymbol{F} &= \boldsymbol{P} + \boldsymbol{\chi}_a^T \boldsymbol{u} \\ \boldsymbol{V}_s &= \boldsymbol{\gamma}^{-1} \boldsymbol{\chi}_s \{\delta\} \end{aligned} \tag{2.83}$$

$\boldsymbol{u}$ is the voltage generated by the piezoelectric sheet, and $\boldsymbol{V}_s$ is the voltage sensed by the piezoelectric sheet.

2.3.2.3 System Expressed in Complex Space

The system equations are casted into the state-space format as follows

$$\begin{aligned} \boldsymbol{A}_\delta \dot{\boldsymbol{x}}_\delta + \boldsymbol{B}_\delta \boldsymbol{x}_\delta &= \boldsymbol{P}_\delta + \boldsymbol{Q}_\delta + \begin{bmatrix} \boldsymbol{0} \\ \boldsymbol{\chi}_a^T \end{bmatrix} \boldsymbol{u} \\ \boldsymbol{V}_s &= \begin{bmatrix} \boldsymbol{0} & \boldsymbol{\gamma}^{-1} \boldsymbol{\chi}_s \end{bmatrix} \boldsymbol{x}_\delta \end{aligned} \tag{2.84}$$

Substituting Eq. (2.61) into the above formula and considering Eq. (2.59), the system equation is

$$\begin{aligned} \dot{\boldsymbol{x}}_q &= \boldsymbol{\lambda} \boldsymbol{x}_q + \boldsymbol{\Pi}\left(\boldsymbol{P}_\delta + \boldsymbol{Q}_\delta\right) + \boldsymbol{\Pi} \begin{bmatrix} \boldsymbol{0} \\ \boldsymbol{\chi}_a^T \end{bmatrix} \boldsymbol{u} \\ \boldsymbol{V}_s &= \begin{bmatrix} \boldsymbol{0} & \boldsymbol{\gamma}^{-1} \boldsymbol{\chi}_s \end{bmatrix} \boldsymbol{\Phi} \boldsymbol{x}_q \end{aligned} \tag{2.85}$$

Only the first $2n$ modes $\boldsymbol{\Phi}_c$ is retained. Then the system can be written as

$$\begin{aligned} \dot{\boldsymbol{x}} &= \boldsymbol{A}\boldsymbol{x} + \boldsymbol{B}\boldsymbol{u} + \boldsymbol{d} \\ \boldsymbol{V}_s &= \boldsymbol{C}\boldsymbol{x} \end{aligned} \tag{2.86}$$

where

$$\begin{aligned} &\boldsymbol{x} = \begin{bmatrix} \boldsymbol{x}_{1\sim n} & \\ & \boldsymbol{x}_{1\sim n}^* \end{bmatrix} \quad \boldsymbol{A} = \begin{bmatrix} \boldsymbol{\lambda}_{1\sim n} & \\ & \boldsymbol{\lambda}_{1\sim n}^* \end{bmatrix} \quad \boldsymbol{\Pi}_c = \begin{bmatrix} \boldsymbol{\Pi}_{1\sim n} & \\ & \boldsymbol{\Pi}_{1\sim n}^* \end{bmatrix} \\ &\boldsymbol{d} = \boldsymbol{\Pi}_c\left(\boldsymbol{P}_\delta + \boldsymbol{Q}_\delta\right) \quad \boldsymbol{B} = \begin{bmatrix} \boldsymbol{B}_{1\sim n} & \\ & \boldsymbol{B}_{1\sim n}^* \end{bmatrix} = \boldsymbol{\Pi}_c \begin{bmatrix} \boldsymbol{0} \\ \boldsymbol{\chi}_a^T \end{bmatrix} \quad \boldsymbol{C} = \begin{bmatrix} \boldsymbol{0} & \boldsymbol{\gamma}^{-1} \boldsymbol{\chi}_s \end{bmatrix} \boldsymbol{\Phi}_c \end{aligned} \tag{2.87}$$

2.3.3 Modeling of Spinning Solar Wing

This section takes the simplified model of the beam-plate combined structure of the solar wing as the object, and establishes the dynamic model of the solar wing with a rotating motion based on the finite element method.

2.3.3.1 Coordinates Definition

In order to describe the motion of the rotating solar wing under the cantilever condition, three coordinate systems are defined, as shown in Fig. 2.12:

(1) Inertial coordinate system *OXYZ*: The coordinate axes are fixed in space, and the directions remain unchanged.
(2) Body coordinate system $o_s x_s y_s z_s$: The coordinate system rotates with the rotation of the solar wing and the origin o_s is fixed at the root of the yoke. Directions of the axis x_s and the axis X are always consistent. The solar wing rotates about the axis x_s, and the angle between the y axis and the Y axis is represented by φ.
(3) Local coordinate system $o_{e,i} x_{e,i} y_{e,i} z_{e,i}$: The coordinate system is connected to the element. The direction of the axis $z_{e,i}$ is consistent with the axis z_s. For the element on the solar panel, the axis $x_{e,i}$ and axis $y_{e,i}$ are parallel to axis x_s and axis y_s, respectively.

Figure 2.13 defines the local coordinate system of the element on the yoke. The local coordinate systems of the elements on the three rods are $o_{e,1} x_{e,1} y_{e,1} z_{e,1}$, $o_{e,2} x_{e,2} y_{e,2} z_{e,2}$ and $o_{e,3} x_{e,3} y_{e,3} z_{e,3}$. Axis $z_{e,1}$, $z_{e,2}$, $z_{e,3}$ are consistent with the axis z_s.

The transformation matrix from body coordinate system to inertial coordinate system is

$$\boldsymbol{T}_{sb} = \begin{bmatrix} 1 & & \\ & c\varphi & -s\varphi \\ & s\varphi & c\varphi \end{bmatrix} \tag{2.88}$$

where $c(\bullet)$represents $cos(\bullet)$ and $s(\bullet)$ represents $sin(\bullet)$.

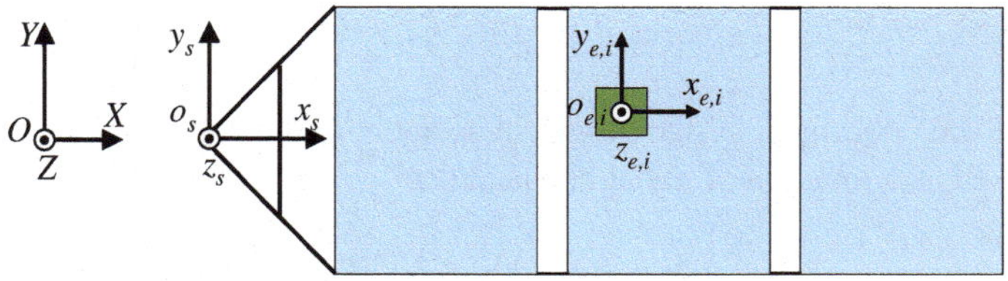

Fig. 2.12 Schematic diagram of the solar wing coordinate system

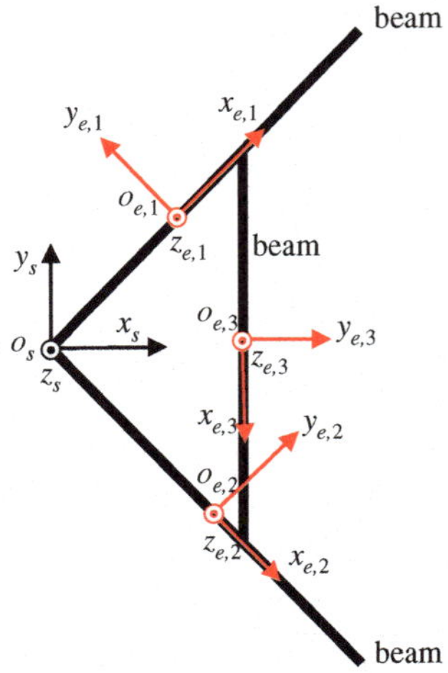

Fig. 2.13 Schematic diagram of the local coordinate systems of the yoke

For elements in the solar panel, the coordinate axes of the local coordinate system of the element are in the same direction as the axes of the body coordinate system, so the transformation matrix is the identity matrix. For the element in the yoke, the local coordinate system of the element has a certain angle with the body coordinate system. Define α as the rotation angle between the $y_{e,i}$ axis and the y_s axis. Then the transformation matrix from the local coordinate system of the element to the body coordinate system is

$$\boldsymbol{T}_{be} = \begin{bmatrix} c\alpha & -s\alpha & \\ s\alpha & c\alpha & \\ & & 1 \end{bmatrix} \tag{2.89}$$

2.3.3.2 Discretization of Flexible Beam and Plate

For the beam element in the yoke, the element has two nodes and each node has six degrees of freedom. So, there are twelve degrees of freedom in total for one element, namely

$$\boldsymbol{\delta}^{e,beam} = \begin{bmatrix} u_i & v_i & w_i & \theta_{xi} & \theta_{yi} & \theta_{zi} & u_j & v_j & w_j & \theta_{xj} & \theta_{yj} & \theta_{zj} \end{bmatrix}^T \tag{2.90}$$

In the local coordinate system, the displacement of any point in the element can be expressed as a function of nodal displacement

$$\boldsymbol{u}_{e,beam} = \boldsymbol{N}_{beam}\boldsymbol{\delta}^{e,beam} \tag{2.91}$$

where $\boldsymbol{N}_{beam}$ is the shape function matrix of the entire element, and the beam does not consider the warping effect. Therefore, the nodal displacement vector can be expressed as

$$\boldsymbol{N}_{beam} = \begin{bmatrix} \boldsymbol{N}_u & & & \\ & \boldsymbol{N}_v & & \\ & & \boldsymbol{N}_w & \\ & & & \boldsymbol{N}_u \\ & & -\boldsymbol{N}'_w & \\ & \boldsymbol{N}'_v & & \end{bmatrix} \tag{2.92}$$

The expression of each matrix in the above equation is shown in Eq. (2.32).

Transform the displacement of the element to the body coordinate system

$$\boldsymbol{u}_{b,beam} = \boldsymbol{T}_{be}\boldsymbol{u}_{e,beam} \tag{2.93}$$

The solar panel is divided by rectangular plate elements, each element has a length of $2a$ and a width of $2b$, as shown in Fig. 2.14. Each node of the plate element has six degrees of freedom. The symbols of the four nodes of element are i, j, k, and l. The displacement of all nodes of an element can be expressed as

$$\begin{aligned} \boldsymbol{\delta}^{e,plate} = [&u_i\ v_i\ w_i\ \theta_{xi}\ \theta_{yi}\ \theta_{zi}\ u_j\ v_j\ w_j\ \theta_{xj}\ \theta_{yj}\ \theta_{zj} \cdots \\ &u_k\ v_k\ w_k\ \theta_{xk}\ \theta_{yk}\ \theta_{zk}\ u_l\ v_l\ w_l\ \theta_{xl}\ \theta_{yl}\ \theta_{zl}]^T \end{aligned} \tag{2.94}$$

The expression of the elastic displacement at any point (x_e,y_e) in the element is

$$\boldsymbol{u}_{e,plate} = \boldsymbol{N}_{plate}\boldsymbol{\delta}^{e,plate} \tag{2.95}$$

$\boldsymbol{N}_p$ is the plate element shape function, and the expression is

$$\boldsymbol{N}_{plate} = \begin{bmatrix} \boldsymbol{N}_x \\ \boldsymbol{N}_y \\ \boldsymbol{N}_z \end{bmatrix} \tag{2.96}$$

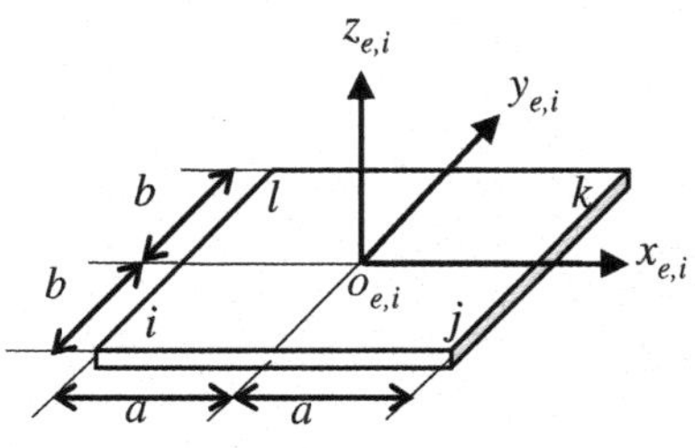

Fig. 2.14 Schematic diagram of local coordinate system of plate element

where

$$\begin{aligned}
\boldsymbol{N}_x &= [\boldsymbol{N}_k \ 0\ 0\ 0\ 0\ 0\ \boldsymbol{N}_l \ 0\ 0\ 0\ 0\ 0\ \boldsymbol{N}_m \ 0\ 0\ 0\ 0\ 0\ \boldsymbol{N}_n \ 0\ 0\ 0\ 0\ 0] \\
\boldsymbol{N}_y &= [0\ \boldsymbol{N}_k \ 0\ 0\ 0\ 0\ 0\ \boldsymbol{N}_l \ 0\ 0\ 0\ 0\ 0\ \boldsymbol{N}_m \ 0\ 0\ 0\ 0\ 0\ \boldsymbol{N}_n \ 0\ 0\ 0\ 0] \\
\boldsymbol{N}_z &= [0\ 0\ \boldsymbol{N}_1 \ 0\ 0\ 0\ \boldsymbol{N}_2 \ 0\ 0\ 0\ \boldsymbol{N}_3\ 0\ 0\ 0\ \boldsymbol{N}_4\ 0]
\end{aligned} \tag{2.97}$$

and

$$\begin{aligned}
\boldsymbol{N}_k &= \frac{1}{4}X_1Y_1 \quad \boldsymbol{N}_l = \frac{1}{4}X_2Y_1 \quad \boldsymbol{N}_m = \frac{1}{4}X_2Y_2 \quad \boldsymbol{N}_n = \frac{1}{4}X_1Y_2 \\
\boldsymbol{N}_1 &= \frac{1}{16}X_1Y_1[X_1Y_1 - X_2Y_2 + 2X_1X_2 + 2Y_1Y_2 \quad 2bY_1Y_2 \quad -2aX_1X_2] \\
\boldsymbol{N}_2 &= \frac{1}{16}X_2Y_1[X_2Y_1 - X_1Y_2 + 2X_1X_2 + 2Y_1Y_2 \quad 2bY_1Y_2 \quad 2aX_1X_2] \\
\boldsymbol{N}_3 &= \frac{1}{16}X_2Y_2[-X_1Y_1 + X_2Y_2 + 2X_1X_2 + 2Y_1Y_2 \quad -2bY_1Y_2 \quad 2aX_1X_2] \\
\boldsymbol{N}_4 &= \frac{1}{16}X_1Y_2[X_1Y_2 - X_2Y_1 + 2X_1X_2 + 2Y_1Y_2 \quad -2bY_1Y_2 \quad -2aX_1X_2]
\end{aligned} \tag{2.98}$$

X_1, X_2, Y_1, Y_2 are intermediate variables, and the expressions are

$$X_1 = 1 - \frac{x_e}{a}; \quad X_2 = 1 + \frac{x_e}{a}; \quad Y_1 = 1 - \frac{y_e}{b}; \quad Y_2 = 1 + \frac{y_e}{b} \tag{2.99}$$

Transform the displacement of the plate element to the body coordinate system

$$\boldsymbol{u}_{b,plate} = \boldsymbol{T}_{be}\boldsymbol{u}_{e,plate} \tag{2.100}$$

2.3.3.3 Energies Formulation

The position vector for an arbitrary point in the i-th element of the beam or plate element can be expressed as

$$\boldsymbol{R}_{s,i} = \boldsymbol{r}_{s,i} + \boldsymbol{u}_i \tag{2.101}$$

In the equation, $\boldsymbol{r}_{s,i}$ is the vector from o_s to o_i, which is given in the body coordinate system. $\boldsymbol{u}_i$ is the elastic displacement vector, which is represented as $\boldsymbol{u}_{b,beam}$ and $\boldsymbol{u}_{b,plate}$ for beam element and plate element, respectively.

The velocity vector for an arbitrary in the i-th element is expressed in the body coordinate system

$$\boldsymbol{V}_{s,i} = \frac{d\boldsymbol{R}_{s,i}}{dt} = \left(\tilde{\boldsymbol{r}}_{s,i} + \tilde{\boldsymbol{u}}\right)^T \boldsymbol{\omega}_f + \dot{\boldsymbol{u}}_i \tag{2.102}$$

$\boldsymbol{\omega}_f$ is the angular velocity vector of the $o_s x_s y_s z_s$ coordinate system relative to the *OXYZ* coordinate system. ' ~ ' indicates the operation of skew matrix.

The kinetic energy of the system is

$$\begin{aligned} T_s &= \frac{1}{2}\sum_{i=1}^{N}\int_{\Omega_i} \boldsymbol{V}_{s,i}^T \boldsymbol{V}_{s,i} dm \\ &= \frac{1}{2}\dot{\boldsymbol{\delta}}^T \boldsymbol{M}_s \dot{\boldsymbol{\delta}} - \boldsymbol{\delta}^T \boldsymbol{G}_7 \dot{\boldsymbol{\delta}} + \frac{1}{2}\boldsymbol{\delta}^T \boldsymbol{G}_8 \boldsymbol{\delta} + \boldsymbol{G}_9^T \boldsymbol{\delta} - \boldsymbol{G}_{10}\dot{\boldsymbol{\delta}} + \frac{1}{2}\boldsymbol{G}_0 \end{aligned} \tag{2.103}$$

$\boldsymbol{\delta}$ is the vector of nodal degrees of freedom, and $\boldsymbol{M}_s$ is the mass matrix. The variables in the above equation are

$$\begin{aligned} &\boldsymbol{M}_s = \sum_{i=1}^{N} \boldsymbol{T}_{eb,i}^T \left(\int_{\Omega_i} \boldsymbol{N}_i^T \boldsymbol{N}_i dm \right) \boldsymbol{T}_{eb,i} && \boldsymbol{G}_7 = \sum_{i=1}^{N} \boldsymbol{T}_{eb,i}^T \left(\int_{\Omega_i} \boldsymbol{N}_i^T \tilde{\boldsymbol{\omega}}_f \boldsymbol{N}_i dm \right) \boldsymbol{T}_{eb,i} \\ &\boldsymbol{G}_8 = \sum_{i=1}^{N} \boldsymbol{T}_{eb,i}^T \left(\int_{\Omega_i} \boldsymbol{N}_i^T \tilde{\boldsymbol{\omega}}_f \tilde{\boldsymbol{\omega}}_f^T \boldsymbol{N}_i dm \right) \boldsymbol{T}_{eb,i} && \boldsymbol{G}_9 = \sum_{i=1}^{N} \boldsymbol{T}_{eb,i}^T \left(\int_{\Omega_i} \boldsymbol{N}_i^T \tilde{\boldsymbol{\omega}}_f \tilde{\boldsymbol{\omega}}_f^T \tilde{\boldsymbol{r}}_{s,i} dm \right) \\ &\boldsymbol{G}_{10} = \sum_{i=1}^{N} \left(\int_{\Omega_i} \tilde{\boldsymbol{r}}_{s,i}^T \tilde{\boldsymbol{\omega}}_f \boldsymbol{N}_i dm \right) \boldsymbol{T}_{eb,i} && \boldsymbol{G}_0 = \sum_{i=1}^{N} \int_{\Omega_i} \tilde{\boldsymbol{r}}_{s,i}^T \boldsymbol{T}_{es,i} \tilde{\boldsymbol{\omega}}_f \tilde{\boldsymbol{\omega}}_f^T \boldsymbol{T}_{es,i}^T \tilde{\boldsymbol{r}}_{s,i} dm \end{aligned} \tag{2.104}$$

The potential energy of the system is

$$U_s = \sum_{i=1}^{N} \int_{\Omega_i} \left(\boldsymbol{\delta}^{e,i}\right)^T \boldsymbol{B}_i^T \boldsymbol{D}_i \boldsymbol{B}_i \boldsymbol{\delta}^{e,i} d\Omega = \boldsymbol{\delta}^T \boldsymbol{K}_s \boldsymbol{\delta} \tag{2.105}$$

where $\boldsymbol{K}_s$ is the stiffness matrix, which can be written as

$$\boldsymbol{K}_s = \sum_{i=1}^{N} \int_{\Omega_i} \boldsymbol{B}_i^T \boldsymbol{D}_i \boldsymbol{B}_i d\Omega \tag{2.106}$$

$\boldsymbol{D}_i$ is the elastic matrix of the *i*-th element, $\boldsymbol{B}_i$ is the matrix related to the shape function $\boldsymbol{N}_i$ [24].

2.3.3.4 System Equation

According to Hamilton's principle, the governing equation of the system can be obtained

$$\boldsymbol{M}_s\ddot{\boldsymbol{\delta}} + \left(\boldsymbol{G}_7 - \boldsymbol{G}_7^T\right)\dot{\boldsymbol{\delta}} + \left[\boldsymbol{K}_s - \left(\boldsymbol{G}_8 + \dot{\boldsymbol{G}}_7^T\right)\right]\boldsymbol{\delta} - \left(\boldsymbol{G}_9 + \dot{\boldsymbol{G}}_{10}^T\right) = 0 \quad (2.107)$$

When the solar wing rotates at a constant rate, the equation is simplified to be

$$\boldsymbol{M}_s\ddot{\boldsymbol{\delta}} + \left(\boldsymbol{G}_7 - \boldsymbol{G}_7^T\right)\dot{\boldsymbol{\delta}} + (\boldsymbol{K}_s - \boldsymbol{G}_8)\boldsymbol{\delta} - \boldsymbol{G}_9 = 0 \quad (2.108)$$

It can be seen from the above equation that after considering the rotation effect, the equivalent stiffness matrix $\boldsymbol{K}(\boldsymbol{K} = \boldsymbol{K}_s\text{-}\boldsymbol{G}_8)$ of the solar wing decreases. The rotation of the solar wing around the root introduces a gyro term $\left(\boldsymbol{G}_7 - \boldsymbol{G}_7^T\right)$, which is an antisymmetric matrix. The natural frequency and mode shape of the structure can be obtained according to the method in Sect. 2.3.1.3.

2.4 Mathematical Model for Flexible Spacecraft with Articulated Appendages[1]

This section presents the formulation of the flexible Spacecraft with articulated appendages using Lagrange's equations in terms of quasi-coordinates. The rigid-flexible coupled equations include the orbital motion, attitude motion, and vibration of the flexible appendages.

2.4.1 Definition of Coordinates

To derive the mathematical model of the system, the hybrid coordinate system is used to describe the motion of the rigid central body and the flexible appendage. A system at reference frames is defined as shown in Fig. 2.15.

2.4.2 Discretization of Flexible Appendage

To describe the elastic deformation of the appendage, the FEM is used to discretize the flexible appendage. When the system is actuated by inner or external loads, the flexible appendage undergoes elastic translations. Let the displacements $\boldsymbol{u}_i$ at any point within the i-th element be approximated as a column vector,

$$\boldsymbol{u}_i = \boldsymbol{N}_i\boldsymbol{\delta}'^{e,i} \quad (2.109)$$

[1] Jie Wang, Dongxu Li, Jianping Jiang. First order coupled dynamic model of flexible space structures with time-varying configurations, Acta Astronautic, 2017,132, 117–123.

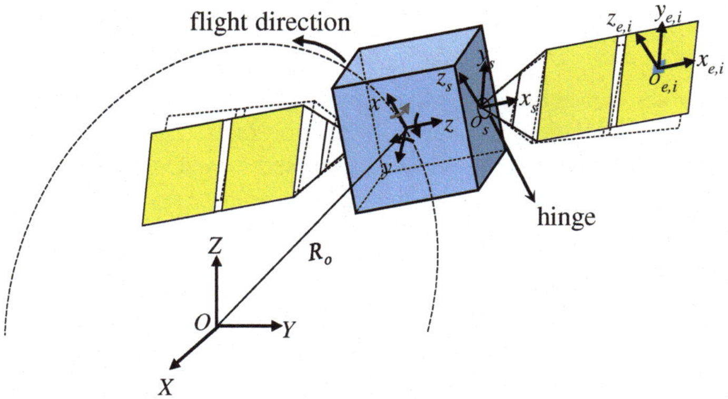

Fig. 2.15 Mathematical model of the spacecraft with flexible articulated appendage

where the function $\boldsymbol{N}_i$ is called shape function and $\boldsymbol{\delta}'^{e,i}$ is nodal freedoms of the element. For different element shapes, the shape function has different choices. The vector $\boldsymbol{u}_i$ and $\boldsymbol{\delta}'^{e,i}$ are both expressed in terms of components along the element axes. Let $\boldsymbol{\delta}^{e,i}$ denotes nodal freedoms in appendage's frame, then the relationship with $\boldsymbol{\delta}'^{e,i}$ can be expressed as

$$\boldsymbol{\delta}'^{e,i} = \begin{bmatrix} \boldsymbol{T}_{es,i} & & \\ & \boldsymbol{T}_{es,i} & \\ & & \ddots \end{bmatrix} \boldsymbol{\delta}^{e,i} = \boldsymbol{T}_{e,i}\boldsymbol{\delta}^{e,i} \tag{2.110}$$

where $\boldsymbol{T}_{es,i}$ denotes the rotation matrix from appendage's frame to the i-th element's local coordinate frame. $\boldsymbol{T}_{e,i}$ represents the transformation matrix of the i-th element. Substitute Eq. (2.110) into Eq. (2.109), the displacement is

$$\boldsymbol{u}_i = \boldsymbol{N}_i \boldsymbol{T}_{e,i}\boldsymbol{\delta}^{e,i} \tag{2.111}$$

2.4.3 Dynamics

The position vector of an arbitrary point P in the rigid body can be written as

$$\boldsymbol{R}_p = \boldsymbol{R}_o + \boldsymbol{r} \tag{2.112}$$

where $\boldsymbol{R}_o$ is the radius vector from O to o, $\boldsymbol{r}$ is the position vector of a point in the rigid body. We note that vectors $\boldsymbol{R}_o$ and $\boldsymbol{r}$ are given in terms of components along xyz. The position vector of a point in the ith element flexible appendage can be written as

$$\boldsymbol{R}_{s,i} = \boldsymbol{T}_{eb,i}(\boldsymbol{R}_o + \boldsymbol{r}_{os}) + \boldsymbol{r}_{s,i} + \boldsymbol{u}_i \tag{2.113}$$

where $\boldsymbol{r}_{os}$ denotes the vector from o to o_s, and is given in the body frame. $\boldsymbol{r}_{s,i}$ denotes the vector from o_s to o_i, and is given in local coordinate frame of the element. $\boldsymbol{u}_i$ is the elastic displacement vector previously discussed. $\boldsymbol{T}_{eb,i}$ denotes the transformation matrix that defines the orientation of the local coordinate system of the ith element with respect to the body coordinate system

$$\boldsymbol{T}_{eb,i} = \boldsymbol{T}_{es,i}\boldsymbol{T}_{sb} \tag{2.114}$$

where $\boldsymbol{T}_{sb}$ is the transformation matrix from the body frame to appendage floating frame

$$\boldsymbol{T}_{sb} = \begin{bmatrix} c\theta_s c\psi_s - s\varphi_s s\theta_s s\psi_s & c\theta_s s\psi_s + s\varphi_s s\theta_s c\psi_s & -c\varphi_s s\theta_s \\ -c\varphi_s s\psi_s & c\varphi_s c\psi_s & s\varphi_s \\ s\theta_s c\psi_s + s\varphi_s c\theta_s s\psi_s & s\theta_s s\psi_s - s\varphi_s c\theta_s c\psi_s & -c\varphi_s c\theta_s \end{bmatrix} \tag{2.115}$$

where $(\varphi_s, \theta_s, \psi_s)$ denotes the Euler angle between the body frame and the appendage floating frame. The derivative of the matrix $\boldsymbol{T}_{sb}$ with respect to time is

$$\dot{\boldsymbol{T}}_{sb} = \tilde{\boldsymbol{\omega}}_f \boldsymbol{T}_{sb} \tag{2.116}$$

$\boldsymbol{\omega}_f$ is the angular velocity vector of axes $x_s y_s z_s$ relative to axes xyz.

The velocity vector of the point P in the rigid body can be expressed in terms of components along xyz as

$$\boldsymbol{V}_p = \frac{d\boldsymbol{R}_p}{dt} = \boldsymbol{V}_o + \boldsymbol{\omega} \times \boldsymbol{r} = \boldsymbol{V}_o + \tilde{\boldsymbol{r}}^T \boldsymbol{\omega} \tag{2.117}$$

where $\boldsymbol{V}_o$ and $\boldsymbol{\omega}$ are the velocity vector and angular velocity vector of the rigid-body with respect to the inertial frame, respectively and expressed in the body frame. The velocity of a deformed point in the i-th element with respect to the local coordinate frame can be calculated as

$$\boldsymbol{V}_{s,i} = \frac{d\boldsymbol{R}_{s,i}}{dt} = \boldsymbol{T}_{eb,i}\left(\boldsymbol{V}_o + \tilde{\boldsymbol{r}}_{os}^T \boldsymbol{\omega}\right) + \left(\tilde{\boldsymbol{r}}_{s,i} + \tilde{\boldsymbol{u}}\right)^T \left(\boldsymbol{T}_{eb,i}\boldsymbol{\omega} + \boldsymbol{T}_{es,i}\boldsymbol{\omega}_f\right) + \dot{\boldsymbol{u}}_i \tag{2.118}$$

Hence, the kinetic energy of the rigid body can be written as

$$T_r = \frac{1}{2}\int_D \boldsymbol{V}_p^T \boldsymbol{V}_p dm = \frac{1}{2} m_r \boldsymbol{V}_o^T \boldsymbol{V}_o + \frac{1}{2} \boldsymbol{\omega}^T \boldsymbol{J}_r \boldsymbol{\omega} \tag{2.119}$$

where m_r denotes the mass of the rigid body and $\boldsymbol{J}_r$ is a symmetric matrix of mass moments of inertia of the rigid body. The superscript T indicates the transposed matrices or vectors.

The kinetic energy of the flexible appendage can be written as

$$\begin{aligned}
T_s &= \frac{1}{2}\sum_{i=1}^{N}\int_{\Omega_i} \boldsymbol{V}_{s,i}^T \boldsymbol{V}_{s,i} dm \\
&= \frac{1}{2} m_s \boldsymbol{V}_o^T \boldsymbol{V}_o + \frac{1}{2}\boldsymbol{\omega}^T \boldsymbol{J}_s \boldsymbol{\omega} + \boldsymbol{V}_o^T \boldsymbol{S}_s^T \boldsymbol{\omega} + \frac{1}{2}\dot{\boldsymbol{\delta}}^T \boldsymbol{M}_s \dot{\boldsymbol{\delta}} + \boldsymbol{V}_o^T \boldsymbol{C}_s \dot{\boldsymbol{\delta}} + \boldsymbol{\omega}^T \boldsymbol{D}_s \dot{\boldsymbol{\delta}} \\
&\quad - \boldsymbol{G}_5^T \boldsymbol{T}_{sb} \boldsymbol{V}_o - \boldsymbol{V}_o^T \boldsymbol{T}_{sb}^T \boldsymbol{G}_3 \boldsymbol{\delta} - \left(\boldsymbol{G}_5^T \boldsymbol{T}_{sb} \tilde{\boldsymbol{r}}_{os}^T + \boldsymbol{G}_6^T \boldsymbol{T}_{sb}\right)\boldsymbol{\omega} - \boldsymbol{\omega}^T \left(\tilde{\boldsymbol{r}}_{os} \boldsymbol{T}_{sb}^T \boldsymbol{G}_3 + \boldsymbol{T}_{sb}^T \boldsymbol{G}_4\right)\boldsymbol{\delta} \\
&\quad - \boldsymbol{\delta}^T \boldsymbol{G}_7 \dot{\boldsymbol{\delta}} + \frac{1}{2}\boldsymbol{\delta}^T \boldsymbol{G}_8 \boldsymbol{\delta} + \boldsymbol{G}_9^T \boldsymbol{\delta} - \boldsymbol{G}_{10} \dot{\boldsymbol{\delta}} + \frac{1}{2}\boldsymbol{G}_0
\end{aligned} \tag{2.120}$$

where $\boldsymbol{\delta}$ is the assembled nodal displacement vector of the appendage, m_s denotes the total mass of the appendage, $\boldsymbol{J}_s$ is the matrix of mass moments of inertia of the appendage, $\boldsymbol{S}_s$ is recognized as a skew-symmetric matrix of first moments of inertia, $\boldsymbol{M}_s$ is the mass matrix of the appendage. When the flexible appendage is fixed on the platform, only the terms with double underlines in the above equation exist. In this case, $\boldsymbol{C}_s$ is denoted as the rigid-flex translational coupling matrix and $\boldsymbol{D}_s$ is the rigid-flex rotational coupling matrix. And they can be expressed as follows

$$\begin{aligned}
\boldsymbol{J}_s &= \sum_{i=1}^{N}\int_{\Omega_i}\left(\boldsymbol{T}_{eb,i}\tilde{\boldsymbol{r}}_{os}^T + \tilde{\boldsymbol{r}}_{s,i}^T \boldsymbol{T}_{eb,i} + \tilde{\boldsymbol{u}}_i^T \boldsymbol{T}_{eb,i}\right)^T \left(\boldsymbol{T}_{eb,i}\tilde{\boldsymbol{r}}_{os}^T + \tilde{\boldsymbol{r}}_{s,i}^T \boldsymbol{T}_{eb,i} + \tilde{\boldsymbol{u}}_i^T \boldsymbol{T}_{eb,i}\right) dm \\
\boldsymbol{S}_s &= \sum_{i=1}^{N}\int_{\Omega_i}\left(\tilde{\boldsymbol{r}}_{os} + \boldsymbol{T}_{eb,i}^T \tilde{\boldsymbol{r}}_{s,i} \boldsymbol{T}_{eb,i} + \boldsymbol{T}_{eb,i}^T \tilde{\boldsymbol{u}}_i \boldsymbol{T}_{eb,i}\right) dm \\
\boldsymbol{M}_s &= \sum_{i=1}^{N} \boldsymbol{T}_{e,i}^T \left(\int_{\Omega_i} \boldsymbol{N}_i^T \boldsymbol{N}_i dm\right) \boldsymbol{T}_{e,i} \\
\boldsymbol{C}_s &= \sum_{i=1}^{N} \boldsymbol{T}_{eb,i}^T \left(\int_{\Omega_i} \boldsymbol{N}_i dm\right) \boldsymbol{T}_{e,i} \\
\boldsymbol{D}_s &= \sum_{i=1}^{N}\left[\int_{\Omega_i}\left(\tilde{\boldsymbol{r}}_{os} \boldsymbol{T}_{eb,i}^T + \boldsymbol{T}_{eb,i}^T \tilde{\boldsymbol{r}}_s + \boldsymbol{T}_{eb,i}^T \tilde{\boldsymbol{u}}_i\right) \boldsymbol{N} dm\right] \boldsymbol{T}_{e,i}
\end{aligned} \tag{2.121}$$

When we assume small elastic motions, terms of order higher than two involving the elastic displacements have been neglected, then

$$\begin{aligned}
\boldsymbol{J}_s &= m_s \tilde{\boldsymbol{r}}_{os} \tilde{\boldsymbol{r}}_{os}^T + \tilde{\boldsymbol{r}}_{os} \boldsymbol{T}_{sb}^T \tilde{\boldsymbol{S}}_e \boldsymbol{T}_{sb} + \left(\tilde{\boldsymbol{r}}_{os} \boldsymbol{T}_{sb}^T \tilde{\boldsymbol{S}}_e \boldsymbol{T}_{sb}\right)^T + \boldsymbol{T}_{sb}^T \boldsymbol{J}_e \boldsymbol{T}_{sb} \\
\boldsymbol{S}_s &= m_s \tilde{\boldsymbol{r}}_{os} + \boldsymbol{T}_{sb}^T \tilde{\boldsymbol{S}}_e \boldsymbol{T}_{sb} \\
\boldsymbol{C}_s &= \boldsymbol{T}_{sb}^T \boldsymbol{G}_1
\end{aligned}$$

$$\boldsymbol{D}_s = \tilde{\boldsymbol{r}}_{os}\boldsymbol{T}_{sb}^T\boldsymbol{G}_1 + \boldsymbol{T}_{sb}^T\boldsymbol{G}_2 \tag{2.122}$$

The various quantities in Eqs. (2.120, 2.121 and 2.122) are

$$\begin{aligned}
&\tilde{\boldsymbol{S}}_e = \sum_{i=1}^{N}\boldsymbol{T}_{es,i}^T\left(\int_{\Omega_i}\tilde{\boldsymbol{r}}_{s,i}dm\right)\boldsymbol{T}_{es,i} && \boldsymbol{J}_e = \sum_{i=1}^{N}\boldsymbol{T}_{es,i}^T\left(\int_{\Omega_i}\tilde{\boldsymbol{r}}_{s,i}\tilde{\boldsymbol{r}}_{s,i}^T dm\right)\boldsymbol{T}_{es,i}\\
&\boldsymbol{G}_1 = \sum_{i=1}^{N}\boldsymbol{T}_{es,i}^T\left(\int_{\Omega_i}\boldsymbol{N}_i dm\right)\boldsymbol{T}_{e,i} && \boldsymbol{G}_2 = \sum_{i=1}^{N}\boldsymbol{T}_{es,i}^T\left(\int_{\Omega_i}\tilde{\boldsymbol{r}}_{s,i}\boldsymbol{N}_i dm\right)\boldsymbol{T}_{e,i}\\
&\boldsymbol{G}_3 = \sum_{i=1}^{N}\boldsymbol{T}_{es,i}^T\left(\int_{\Omega_i}\tilde{\boldsymbol{\omega}}_f^T\boldsymbol{N}_i dm\right)\boldsymbol{T}_{e,i} && \boldsymbol{G}_4 = \sum_{i=1}^{N}\boldsymbol{T}_{es,i}^T\left(\int_{\Omega_i}\tilde{\boldsymbol{r}}_{s,i}\tilde{\boldsymbol{\omega}}_f^T\boldsymbol{N}_i dm\right)\boldsymbol{T}_{e,i}\\
&\boldsymbol{G}_5 = \sum_{i=1}^{N}\boldsymbol{T}_{es,i}^T\left(\int_{\Omega_i}\tilde{\boldsymbol{\omega}}_f^T\tilde{\boldsymbol{r}}_{s,i}dm\right) && \boldsymbol{G}_6 = \sum_{i=1}^{N}\boldsymbol{T}_{es,i}^T\left(\int_{\Omega_i}\tilde{\boldsymbol{r}}_{s,i}^T\tilde{\boldsymbol{\omega}}_f^T\tilde{\boldsymbol{r}}_{s,i}dm\right)\\
&\boldsymbol{G}_7 = \sum_{i=1}^{N}\boldsymbol{T}_{e,i}^T\left(\int_{\Omega_i}\boldsymbol{N}_i^T\tilde{\boldsymbol{\omega}}_f\boldsymbol{N}_i dm\right)\boldsymbol{T}_{e,i} && \boldsymbol{G}_8 = \sum_{i=1}^{N}\boldsymbol{T}_{e,i}^T\left(\int_{\Omega_i}\boldsymbol{N}_i^T\tilde{\boldsymbol{\omega}}_f\tilde{\boldsymbol{\omega}}_f^T\boldsymbol{N}_i dm\right)\boldsymbol{T}_{e,i}\\
&\boldsymbol{G}_9 = \sum_{i=1}^{N}\boldsymbol{T}_{e,i}^T\left(\int_{\Omega_i}\boldsymbol{N}_i^T\tilde{\boldsymbol{\omega}}_f\tilde{\boldsymbol{\omega}}_f^T\tilde{\boldsymbol{r}}_{s,i}dm\right) && \boldsymbol{G}_{10} = \sum_{i=1}^{N}\left(\int_{\Omega_i}\tilde{\boldsymbol{r}}_{s,i}^T\tilde{\boldsymbol{\omega}}_f\boldsymbol{N}_i dm\right)\boldsymbol{T}_{e,i}
\end{aligned}$$

$$\boldsymbol{G}_0 = \sum_{i=1}^{N}\int_{\Omega_i}\tilde{\boldsymbol{r}}_{s,i}^T\boldsymbol{T}_{es,i}\tilde{\boldsymbol{\omega}}_f\tilde{\boldsymbol{\omega}}_f^T\boldsymbol{T}_{es,i}^T\tilde{\boldsymbol{r}}_{s,i}dm \tag{2.123}$$

The potential energy is due to the deformation of the flexible appendage, which has been introduced in the Sect. 2.3.3.3 and can be can be shown to have the expression

$$U_s = \sum_{i=1}^{N}\int_{\Omega_i}\left(\boldsymbol{\delta}^{e,i}\right)^T\boldsymbol{B}_i^T\boldsymbol{D}_i\boldsymbol{B}_i\boldsymbol{\delta}^{e,i}d\Omega = \boldsymbol{\delta}^T\boldsymbol{K}_s\boldsymbol{\delta} \tag{2.124}$$

We express the elastic displacements as linear combinations of space-dependent admissible functions multiplied by time-dependent generalized coordinates, or

$$\boldsymbol{\delta} = \boldsymbol{\Phi}\eta \tag{2.125}$$

where η is a vector of generalized coordinates, $\boldsymbol{\Phi}$ is the shape matrix and satisfy

$$\begin{aligned}
&\boldsymbol{I} = \boldsymbol{\Phi}^T\boldsymbol{M}_s\boldsymbol{\Phi}\\
&\boldsymbol{\Lambda}^2 = \boldsymbol{\Phi}^T\boldsymbol{K}_s\boldsymbol{\Phi} = diag\left(\omega_1^2, \omega_2^2, \ldots, \omega_n^2\right)
\end{aligned} \tag{2.126}$$

where $\boldsymbol{I}$ denotes identity matrix $\boldsymbol{I} \in \mathbb{R}^{n\times n}$. n is the number of freedoms of the appendage. $\omega_1, \omega_2, \ldots\omega_n$ are radial frequencies.

Specify

$$\boldsymbol{C} = \boldsymbol{C}_s\boldsymbol{\Phi} \quad \boldsymbol{D} = \boldsymbol{D}_s\boldsymbol{\Phi} \quad \overline{\boldsymbol{G}}_1 = \boldsymbol{G}_1\boldsymbol{\Phi}$$

$$\overline{\boldsymbol{G}}_2 = \boldsymbol{G}_2\boldsymbol{\Phi} \quad \overline{\boldsymbol{G}}_3 = \boldsymbol{G}_3\boldsymbol{\Phi} \quad \overline{\boldsymbol{G}}_4 = \boldsymbol{G}_4\boldsymbol{\Phi}$$
$$\overline{\boldsymbol{G}}_7 = \boldsymbol{\Phi}^T\boldsymbol{G}_7\boldsymbol{\Phi} \quad \overline{\boldsymbol{G}}_8 = \boldsymbol{\Phi}^T\boldsymbol{G}_8\boldsymbol{\Phi} \quad \overline{\boldsymbol{G}}_9 = \boldsymbol{\Phi}^T\boldsymbol{G}_9$$
$$\overline{\boldsymbol{D}}_2 = -\tilde{\boldsymbol{r}}_{os}\boldsymbol{T}_{sb}^T\overline{\boldsymbol{G}}_3 - \boldsymbol{T}_{sb}^T\overline{\boldsymbol{G}}_4 \quad \boldsymbol{D}_1 = -\tilde{\boldsymbol{r}}_{os}\boldsymbol{T}_{sb}^T\boldsymbol{G}_5 - \boldsymbol{T}_{sb}^T\boldsymbol{G}_6 \tag{2.127}$$

By inserting Eqs. (2.125, 2.126 and 2.127) into Eqs. (2.120) and (2.124), the kinetic and potential energies of the flexible appendage can be written as

$$\begin{aligned} T_s = & \frac{1}{2}m_s\boldsymbol{V}_o^T\boldsymbol{V}_o + \frac{1}{2}\boldsymbol{\omega}^T\boldsymbol{J}_s\boldsymbol{\omega} + \boldsymbol{V}_o^T\boldsymbol{S}_s^T\boldsymbol{\omega} + \frac{1}{2}\dot{\boldsymbol{\eta}}^T\boldsymbol{I}_n\dot{\boldsymbol{\eta}} + \boldsymbol{V}_o^T\boldsymbol{C}\dot{\boldsymbol{\eta}} + \boldsymbol{\omega}^T\boldsymbol{D}\dot{\boldsymbol{\eta}} \\ & - \boldsymbol{G}_5^T\boldsymbol{T}_{sb}\boldsymbol{V}_o - \boldsymbol{V}_o^T\boldsymbol{T}_{sb}^T\overline{\boldsymbol{G}}_3\boldsymbol{\eta} + \boldsymbol{D}_1^T\boldsymbol{\omega} + \boldsymbol{\omega}^T\overline{\boldsymbol{D}}_2\boldsymbol{\eta} \\ & - \boldsymbol{\eta}^T\overline{\boldsymbol{G}}_7\dot{\boldsymbol{\eta}} + \frac{1}{2}\boldsymbol{\eta}^T\overline{\boldsymbol{G}}_8\boldsymbol{\eta} + \boldsymbol{G}_9^T\boldsymbol{\eta} - \boldsymbol{G}_{10}\dot{\boldsymbol{\eta}} + \frac{1}{2}\boldsymbol{G}_0 \end{aligned} \tag{2.128}$$

and

$$U_s = \boldsymbol{\eta}^T\boldsymbol{\Lambda}^2\boldsymbol{\eta} \tag{2.129}$$

2.4.4 Governing Equations

The Lagrange function can be expressed as

$$L = T_r + T_s - U_s \tag{2.130}$$

By using Lagrange's equations in terms of quasi-coordinates [1],

$$\begin{aligned} & \frac{d}{dt}\left(\frac{\partial L}{\partial \boldsymbol{V}}\right) + \tilde{\boldsymbol{\omega}}\left(\frac{\partial L}{\partial \boldsymbol{V}}\right) = \boldsymbol{F}_v \\ & \frac{d}{dt}\left(\frac{\partial L}{\partial \boldsymbol{\omega}}\right) + \tilde{\boldsymbol{V}}\left(\frac{\partial L}{\partial \boldsymbol{V}}\right) + \tilde{\boldsymbol{\omega}}\left(\frac{\partial L}{\partial \boldsymbol{\omega}}\right) = \boldsymbol{M}_\omega \\ & \frac{d}{dt}\left(\frac{\partial L}{\partial \dot{\boldsymbol{q}}}\right) - \frac{\partial L}{\partial \boldsymbol{q}} = 0 \end{aligned} \tag{2.131}$$

Specify $m = m_r + m_s, \boldsymbol{J} = \boldsymbol{J}_r + \boldsymbol{J}_s$, then we obtain equations of motion as follows

$$
\begin{aligned}
\boldsymbol{F}_v &= m\dot{\boldsymbol{V}}_o + \boldsymbol{S}_s^T\dot{\boldsymbol{\omega}} + \underline{\dot{\boldsymbol{S}}_s^T}\boldsymbol{\omega} + \boldsymbol{C}\ddot{\boldsymbol{\eta}} + \left(\underline{\dot{\boldsymbol{C}}} - \boldsymbol{T}_{sb}^T\bar{\boldsymbol{G}}_3\right)\dot{\boldsymbol{\eta}} \\
&\quad - \left(\underline{\dot{\boldsymbol{T}}_{sb}^T}\bar{\boldsymbol{G}}_3 + \underline{\underline{\boldsymbol{T}_{sb}^T\dot{\bar{\boldsymbol{G}}}_3}}\right)\boldsymbol{\eta} - \underline{\dot{\boldsymbol{T}}_{sb}^T}\boldsymbol{G}_5 - \underline{\underline{\boldsymbol{T}_{sb}^T\dot{\boldsymbol{G}}_5}} \\
&\quad + \tilde{\boldsymbol{\omega}}\left(m\boldsymbol{V}_o + \boldsymbol{S}_s^T\boldsymbol{\omega} + \boldsymbol{C}\dot{\boldsymbol{\eta}} - \boldsymbol{T}_{sb}^T\bar{\boldsymbol{G}}_3\boldsymbol{\eta} - \boldsymbol{T}_{sb}^T\boldsymbol{G}_5\right) \\
\boldsymbol{M}_\omega &= \boldsymbol{S}_s\dot{\boldsymbol{V}}_o + \underline{\dot{\boldsymbol{S}}_s}\boldsymbol{V}_o + \boldsymbol{J}\dot{\boldsymbol{\omega}} + \underline{\dot{\boldsymbol{J}}}\boldsymbol{\omega} + \boldsymbol{D}\ddot{\boldsymbol{\eta}} + \underbrace{\left(\bar{\boldsymbol{D}}_2 + \dot{\boldsymbol{D}}\right)\dot{\boldsymbol{\eta}} + \dot{\bar{\boldsymbol{D}}}_2\boldsymbol{\eta} + \dot{\boldsymbol{D}}_1} \\
&\quad + \tilde{\boldsymbol{\omega}}\left(\boldsymbol{J}\boldsymbol{\omega} + \boldsymbol{S}_s\boldsymbol{V}_o + \boldsymbol{D}\dot{\boldsymbol{\eta}} + \bar{\boldsymbol{D}}_2\boldsymbol{\eta} + \boldsymbol{D}_1\right) \\
&\quad + \tilde{\boldsymbol{V}}_o\left(m\boldsymbol{V}_o + \boldsymbol{S}_s^T\boldsymbol{\omega} + \bar{\boldsymbol{C}}_s\dot{\boldsymbol{\eta}} - \boldsymbol{T}_{sb}^T\bar{\boldsymbol{G}}_3\boldsymbol{\eta} - \boldsymbol{T}_{sb}^T\boldsymbol{G}_5\right) \\
0 &= \boldsymbol{C}^T\dot{\boldsymbol{V}}_o + \left(\underline{\dot{\boldsymbol{C}}^T} - \bar{\boldsymbol{G}}_3^T\boldsymbol{T}_{sb}\right)\boldsymbol{V}_o + \boldsymbol{D}^T\dot{\boldsymbol{\omega}} + \left(\underbrace{\dot{\boldsymbol{D}}^T} - \bar{\boldsymbol{D}}_2^T\right)\boldsymbol{\omega} \\
&\quad + \ddot{\boldsymbol{\eta}} + \left(\underbrace{\bar{\boldsymbol{G}}_7 - \bar{\boldsymbol{G}}_7^T}\right)\dot{\boldsymbol{\eta}} + \left(\boldsymbol{\Lambda}^2 - \bar{\boldsymbol{G}}_8 - \underline{\underline{\dot{\bar{\boldsymbol{G}}}_7^T}}\right)\boldsymbol{\eta} - \boldsymbol{G}_9 - \underline{\underline{\dot{\boldsymbol{G}}_{10}^T}}
\end{aligned}
\tag{2.132}
$$

The terms $\boldsymbol{F}_v$ and $\boldsymbol{M}_\omega$ on the left side of the above equation are force and torque vectors applied to the platform, respectively, both presented in terms of components along axes xyz. We assume that the appendage is not activated by external loads, so the left side of the third equation is equal to zero.

In Eq. (2.132), $\dot{\boldsymbol{S}}_s$, $\dot{\boldsymbol{D}}_1$, $\dot{\boldsymbol{C}}$, $\dot{\bar{\boldsymbol{D}}}_2$ are the first order terms induced by the derivative of the coordinate transformation matrix, and have the following forms

$$
\begin{aligned}
\dot{\boldsymbol{S}}_s &= \dot{\boldsymbol{T}}_{sb}^T\tilde{\boldsymbol{S}}_e\boldsymbol{T}_{sb} + \boldsymbol{T}_{sb}^T\tilde{\boldsymbol{S}}_e\dot{\boldsymbol{T}}_{sb} \\
\dot{\boldsymbol{D}}_1 &= -\tilde{\boldsymbol{r}}_{os}\dot{\boldsymbol{T}}_{sb}^T\boldsymbol{G}_5 - \dot{\boldsymbol{T}}_{sb}^T\boldsymbol{G}_6 \\
\dot{\boldsymbol{C}} &= \dot{\boldsymbol{T}}_{sb}^T\boldsymbol{G}_1\boldsymbol{\Phi} \\
\dot{\bar{\boldsymbol{D}}}_2 &= -\tilde{\boldsymbol{r}}_{os}\dot{\boldsymbol{T}}_{sb}^T\bar{\boldsymbol{G}}_3 - \dot{\boldsymbol{T}}_{sb}^T\bar{\boldsymbol{G}}_4
\end{aligned}
\tag{2.133}
$$

The motion of flexible spacecraft is conveniently described by six ordinary differential equations for the three rigid-body translations and rotations of the platform, and by partial differential equations for the elastic deformations of the appendage. In addition, this is no ordinary time-invariable problem, as the maneuvering of the appendage relative to the platform induces time-dependent coefficients.

In Eq. (2.132), terms with single underline contain partial derivatives of the transformation matrix $\boldsymbol{T}_{sb}$, terms with double underlines contain partial derivatives of the angular velocity vector $\boldsymbol{\omega}_f$ and terms with bottom bracket contain both. In previous literature, partial derivatives of terms containing the transformation matrix between axes xyz and $x_sy_sz_s$ with respect to time t are neglected in most cases. When the appendage undergoes fast rotation relative to the platform, these terms may have great influence on the dynamical response of the system, which will be numerically simulated in the next section. The term with bottom wave, which is a skew-symmetric matrix, represents the contribution of the gyroscope effect induced by rotating of the appendage.

When skipping the translational motion of the base and neglecting the first order terms of transformation matrix, Eq. (2.132) has the following form [11, 13]

$$\begin{aligned} \boldsymbol{J}\dot{\boldsymbol{\omega}} + \boldsymbol{D}\ddot{\boldsymbol{\eta}} + \bar{\boldsymbol{D}}_2\dot{\boldsymbol{\eta}} + \tilde{\boldsymbol{\omega}}\left(\boldsymbol{J}\boldsymbol{\omega} + \boldsymbol{D}\dot{\boldsymbol{\eta}} + \bar{\boldsymbol{D}}_2\boldsymbol{\eta} + \boldsymbol{D}_1\right) &= \boldsymbol{M}_\omega \\ \boldsymbol{D}^T\dot{\boldsymbol{\omega}} - \bar{\boldsymbol{D}}_2^T\boldsymbol{\omega} + \ddot{\boldsymbol{\eta}} + \left(\bar{\boldsymbol{G}}_7 - \bar{\boldsymbol{G}}_7^T\right)\dot{\boldsymbol{\eta}} + \left(\boldsymbol{\Lambda}^2 - \bar{\boldsymbol{G}}_8\right)\boldsymbol{\eta} - \boldsymbol{G}_9 &= 0 \end{aligned} \tag{2.134}$$

When the appendage is fixed to the base, Eq. (2.134) has the following form [25–27]

$$\begin{aligned} \boldsymbol{J}\dot{\boldsymbol{\omega}} + \boldsymbol{D}\ddot{\boldsymbol{\eta}} + \tilde{\boldsymbol{\omega}}(\boldsymbol{J}\boldsymbol{\omega} + \boldsymbol{D}\dot{\boldsymbol{\eta}}) &= \boldsymbol{M}_\omega \\ \boldsymbol{D}^T\dot{\boldsymbol{\omega}} + \ddot{\boldsymbol{\eta}} + \boldsymbol{\Lambda}^2\boldsymbol{\eta} &= 0 \end{aligned} \tag{2.135}$$

Generally, when the flexible appendage fixes to the rigid platform of the spacecraft, the inertia matrix $\boldsymbol{J}$ and the coupling matrix $\boldsymbol{D}$ do not change with time. In cases that the flexible appendage moves relative to the body, the moment of inertia of the appendage changes continuously due to the change in relative position, so the moment of inertia of the system varies with respect to time. At the same time, the rotational coupling matrix is also time-varying.

2.5 Summary

The rigid-flexible coupling dynamic models of the appendage and the whole system have been established in this chapter, which is the basis of the subsequent dynamic analysis and control system design.

The main conclusions include:

(1) The configurations of some typical spacecraft have been analyzed. And simplified models for flexible appendages and spacecraft are given.

(2) For flexible appendages with rigid motions, the rigid-flexible coupling dynamic models have been established by the analytical method and the finite element method. It can be seen from the mathematical model of the system that the rigid motion increases the complexity of the system's governing equations, which causes vibrations along the unrelated directions to be coupled with each other.

(3) This chapter has established the first order coupled model for time-varying spacecraft containing rotating flexible appendages by introducing the first order derivative of the transformation matrix. The first order terms altered the rigid-flexible coupling characteristics between the rigid motion of the platform and the elastic deformations of appendages.

References

1. Meirovitch, L., & Kwak, M. K. (1990). Dynamics and control of spacecraft with retargeting flexible antennas. *Journal of Guidance, Control, and Dynamics, 13*(2), 241–248.
2. Schiehlen, W. (1997). Multibody system dynamics: roots and perspectives. *Multibody System Dynamics, 1*, 149–188
3. Shabana, A. A. (1997). Flexible multibody dynamics: review of past and recent developments. *Multibody System Dynamics, 1*(2), 189–222.
4. Santini, P., & Gasbarri, P. (2004). Dynamics of multibody systems in space environment; Lagrangian vs. Eulerian approach. *Acta Astronautica, 54*(1), 1–24.
5. Meirovitch, L., & Nelson, H. (1966). On the high-spin motion of a satellite containing elastic parts. *Journal of Spacecraft and Rockets, 3*(11), 1597–1602.
6. Meirovitch, L. (1991). Hybrid state equations of motion for flexible bodies in terms of quasi-coordinates. *Journal of Guidance, Control, and Dynamics, 14*(5), 1008–1013.
7. Meirovitch, L., & Stemple, T. (1995). Hybrid equations of motion for flexible multibody systems using quasicoordinates. *Journal of Guidance, Control, and Dynamics, 18*(4), 678–688.
8. Alazard, D., Cumer, C., & Tantawi, K. (2008). Linear dynamic modeling of spacecraft with various flexible appendages and on-board angular momentums. In *7th International ESA Conference on Guidance, Navigation & Control Systems*. Tralee, Ireland.
9. Liu, J-Y., & Lu, H. (2007). Rigid-flexible coupling dynamics of three-dimensional hub-beams system. *Multibody System Dynamics, 18*(4), 487–510.
10. Deng, F., He, X., Li, L., et al. (2007). Dynamics modeling for a rigid-flexible coupling system with nonlinear deformation field. *Multibody System Dynamics, 18*(4), 559–578.
11. Gasbarri, P., Monti, R., de Angelis, C., et al. (2014). Effects of uncertainties and flexible dynamic contributions on the control of a spacecraft full-coupled model. *Acta Astronautica, 94*(1), 515–526.
12. Likins, P. W. (1972). Finite element appendage equations for hybrid coordinate dynamic analysis. *International Journal of Solids and Structures, 8*(5), 709–731.
13. Gasbarri, P., Monti, R., & Sabatini, M. (2014). Very large space structures: Non-linear control and robustness to structural uncertainties. *Acta Astronautica*, 93, 252–265.
14. Loquen, T., De Plinval, H., Cumer, C., et al. (2012). Attitude control of satellites with flexible appendages: a structured H_∞ control design. In *Proceedings of the AIAA Guidance, Navigation, and Control Conference*. Minneapolis, USA.
15. Yucelen, T., De la Torre, G., Haddad, W. M., et al. (2013) . Application of a robust adaptive control architecture to a spacecraft with flexible dynamics.
16. Lu, D., & Liu, Y. (2014). Singular formalism and admissible control of spacecraft with rotating flexible solar array. *Chinese Journal of Aeronautics, 27*(1), 136–144.
17. Canavin, J., & Likins, P. (1977). Floating reference frames for flexible spacecraft . *Journal of Spacecraft and Rockets, 14*(12), 724–732.
18. Benson, D., & Hallquist, J. (1986). A simple rigid body algorithm for structural dynamics programs. *International Journal for Numerical Methods in Engineering, 22*(3), 723–749.
19. Banerjee, J., & Su, H. (2004). Development of a dynamic stiffness matrix for free vibration analysis of spinning beams. *Computers & structures, 82*(23), 2189–2197.
20. Tanaka, M., & Bercin, A. (1999). Free vibration solution for uniform beams of nonsymmetrical cross section using Mathematica. *Computers & structures, 71*(1), 1–8.
21. Li, D. (2010). *Advanced dynamics of structures*. Science Press.
22. Wang, Z., & Zhao, J. (2011). Researched torsion theory of open thin-walled beams and its application. *Chinese Journal of Theoretical and Applied Mechanics, 43*(5), 963–967.
23. Li, D. (1985). Some general concepts of complex mode theory. *Journal of Tsinghua University, 25*(3), 26–37.
24. Zienkiewicz, O. C., & Taylor, R. L. (2005). The finite element method for solid and structural mechanics. Butterworth-heinemann.
25. di Gennaro, S. (1998). Adaptive robust tracking for flexible spacecraft in presence of disturbances. *Journal of Optimization Theory and Applications, 98*(3), 545–568.

26. di Gennaro, S. (2003). Passive attitude control of flexible spacecraft from quaternion measurements. *Journal of optimization theory and applications, 116*(1), 41–60.
27. Shahriari, S., Azadi, S., & Moghaddam, M. M. (2010). An accurate and simple model for flexible satellites for three-dimensional studies. *Journal of Mechanical Science and Technology, 24*(6), 1319–1327.
28. Turcic, D. A., & Midha, A. (1984). Dynamic analysis of elastic mechanism systems, Part I: applications. *ASME Journal of Dynamic Systems, Measurment and Control, 106*, 243–248.
29. Likins, P. W. (1972). Finite element appendage equations for hybrid coordinate dynamics analysis. *Journal of Solids & Structures, 8*, 709–731.
30. Ho, J. (1974). The direct path method for deriving the dynamic equations of motion of a multibody flexible spacecraft with topological tree configuration. In *AIAA Paper* (pp. 74–786).
31. Ho, J. Y. L. (1975). Direct path method for flexible multibody spacecraft dynamics. *J SPACECRAFT, 14*(2), 102–110.

Chapter 3
Rigid-Flexible Coupling Dynamic Analysis of Flexible Spacecraft

3.1 Introduction

The main purpose of this chapter is to calculate the dynamic characteristics of the flexible spacecraft with time-varying parameters based on the model in the previous chapter, and to obtain the dynamic responses of the spacecraft under typical cases. This chapter mainly focuses on two aspects: (1) the influence of variable parameters on the dynamic characteristics of flexible appendages; (2) the analysis of dynamic characteristics of the flexible spacecraft, including the influence of variable parameters on system parameters and analysis of the coupling effect of rigid motion and elastic vibration of flexible appendages.

3.2 Dynamic Analysis of Flexible Appendages

In Sect. 2.3, a dynamic model is established for the spinning beam and beam-plate combined structure of the flexible appendage. In this section, numerical methods are used to verify the model, and the influence of structural parameters on dynamical characteristics is analyzed.

© The Author(s), under exclusive license to Springer Nature Singapore Pte Ltd. 2022

J. Wang and D.-X. Li, *Rigid-Flexible Coupling Dynamics and Control of Flexible Spacecraft with Time-Varying Parameters*,
https://doi.org/10.1007/978-981-16-5097-0_3

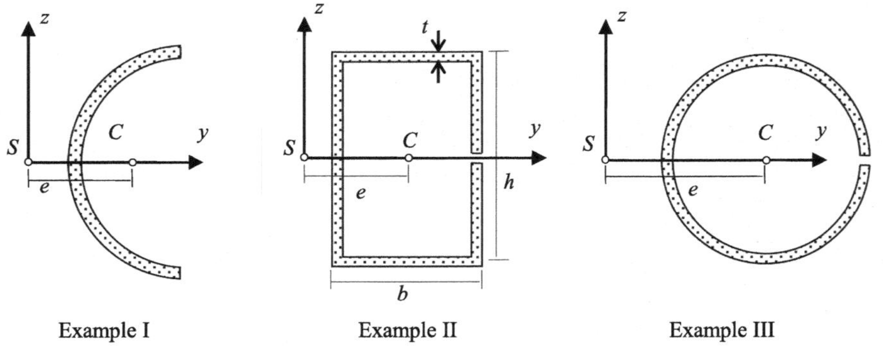

Fig. 3.1 The cross-sections of the three beams studied in the examples

3.2.1 Characteristics of Spinning Beam[1]

This section first uses literature data to verify the correctness of the model, and then analyzes the influence of rotation and warping on modal and frequency.

3.2.1.1 Simulation Cases and Parameters

In order to verify the correctness of the rotation factor in the model, a rectangular cross-section beam is selected for analysis and comparison, and the warping factor is ignored. The simulation example comes from the literature [1]. The parameters are as follows: $EI_{yy} = 582.996\ \text{Nm}^2$, $EI_{zz} = 582.996\ \text{Nm}^2$, $m = 2.87$ kg/m, $L = 1.29$ m. In order to analyze the effect of section warping, three types of beams are selected, as shown in Fig. 3.1. The first is a semicircular opening and thin-wall section. And the second is a square opening and the third is a circular opening. Thin-walled section. The first two are used to verify the model, and the third is used for analysis. Physical properties are shown in Table 3.1 [2].

3.2.1.2 Model Validation

Model verification is divided into two steps. The first step is to verify the rotation effect by comparing the calculated natural frequencies of the rectangular cross-section beam with the results from the literature with ignoring the warping factor. The second step is to verify the warping of the section. The influence of the warping effect on natural frequencies of two beams with unsymmetrical cross-sections are researched and compared with the results in the literature.

[1] Jie Wang, Dongxu Li, Jianping Jiang. Modeling and analysis for coupled flexural–torsional spinning beams with unsymmetrical cross sections. Journal of Theoretical and Applied Mechanics, 2017, 55(1), 213–226.

Table 3.1 Physical properties studied in the examples

	Example I	Example II	Example III
EI_y (Nm^2)	6380	1.436×10^5	185.5
EI_z (Nm^2)	2702	2.367×10^5	1670
GJ (Nm^2)	43.46	346.71	0.022
$E\Gamma$ (Nm^4)	0.10473	536.51	0.057
I_s (kg·m)	0.501×10^{-3}	3.17×10^{-2}	6.739×10^{-5}
m (kg/m)	0.835	4.256	0.133
L (m)	0.82	2.7	5.91
e (m)	0.0155	0.0735	0.0218

(1) Verification of rotation effect

The non-dimensional natural frequency and spinning speed parameter are defined as in the literature [1]

$$\omega_i^* = \omega_i/\omega_0 \qquad \Omega^* = \Omega/\omega_0 \tag{3.1}$$

where

$$\omega_0 = \sqrt{\frac{\sqrt{EI_{yy}I_{zz}}}{\rho AL^4}} \tag{3.2}$$

Comparison of the first three natural frequencies for the current study with those given in published literature is listed in Table 3.2. Both examples apply to cantilever end conditions and the effect of warping stiffness are excluded in the analysis. It is concluded that the resulting frequencies are in good agreement with the one given in previous work. Because EI_{yy} equals EI_{zz} for this example, the natural frequency parameters of the first two modes are equal when the spinning speed parameter is zero.

(2) Verification of warping effect

Table 3.2 Natural frequencies of a spinning beam: (1) reference [1]; (2) present method

Spinning speed parameter (Ω^*)	Natural frequency parameters ($\omega_i{}^*$)					
	$\omega_1{}^*$		$\omega_2{}^*$		$\omega_3{}^*$	
	(1)	(2)	(1)	(2)	(1)	(2)
0	3.516	3.516	3.516	3.516	22.034	22.034
2	1.516	1.516	5.516	5.516	24.034	24.034
3.5	0	0	7.016	7.016	25.534	25.534
4	–	–	7.516	7.516	26.034	26.034

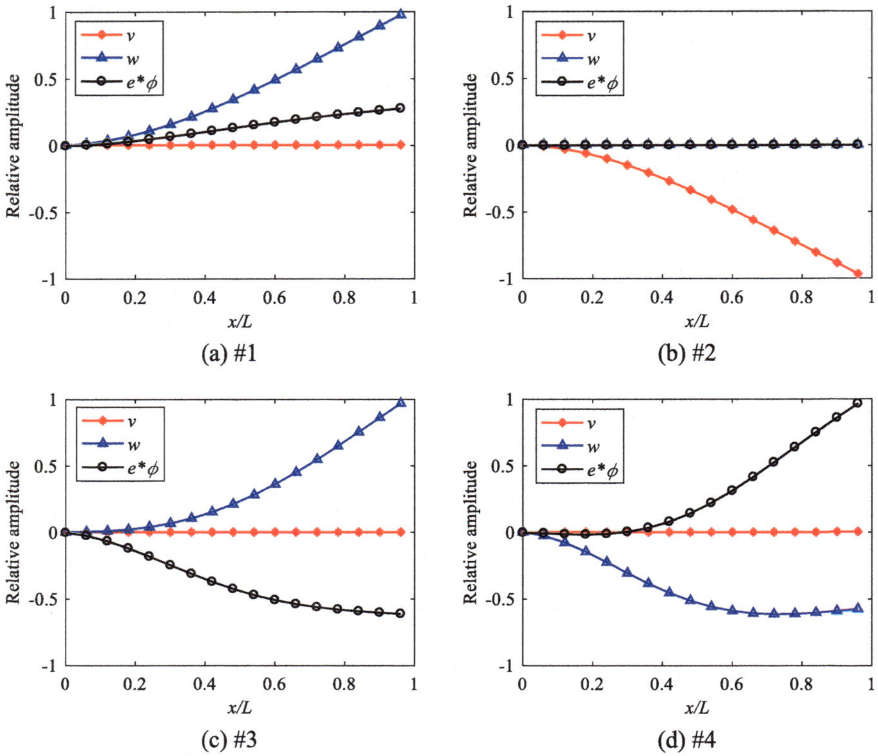

Fig. 3.2 Mode shapes of Example III with speed parameter for $\Omega^* = 0$

Then, to investigate the characteristics of a beam with unsymmetrical cross section, two uniform beams with a semi-circular open cross section and with a channel cross section are considered. The first seven natural frequencies for beams given in Fig. 3.2 are obtained by including and excluding the effect of warping stiffness when the spinning speed is zero, and compared with results from the literature [2], as shown in Table 3.3. The results show that when warping is considered, the results of the method used in this paper are consistent with those in the literature. When neglecting the influence of the warping effect, as the modal order increases, the natural frequency deviation becomes larger. At the same time, when warping is not considered, compared with Example I, the natural frequency deviation of the structure in Example II is larger. This is because the ratio of warping stiffness to structural stiffness in Example II is larger, so the contribution of warping to the natural frequency is greater.

Table 3.3 Natural frequency of rotating beam (Hz): (1) literature [2]; (2) with warping effect; (3) without warping effect

Mode	Example I			Example II		
	(1)	(2)	(3)	(1)	(2)	(3)
#1	63.79	63.79	62.65	11.03	11.02	8.332
#2	137.7	137.7	130.4	–	18.10	18.10
#3	–	149.7	149.7	39.02	39.02	23.92
#4	278.4	278.4	261.5	58.19	58.20	36.74
#5	484.8	484.8	422.5	–	113.4	47.42
#6	663.8	663.8	613.3	152.4	152.4	67.41
#7	–	768.4	656.3	209.4	209.4	86.64

3.2.1.3 Natural Frequencies and Complex Mode Shapes

Table 3.4, Figs. 3.2 and 3.3 respectively show the first four natural frequencies and normalized mode shapes of Example III when $\Omega^* = 0$ and $\Omega^* = 1$. The results show that the rotation along the x-axis changes the natural frequency of the system, especially in the low-order modes, which will be explained in detail in the next section. From Fig. 3.2, the z-direction bending mode (w) and the x-direction torsional mode (ϕ) are always coupled. Since the section is symmetrical about the y-axis, when the dimensionless speed $\Omega^* = 0$, the y-direction bending mode (v) and the x-direction torsional mode are not coupled. When the dimensionless speed $\Omega^* = 1$, the low-order modes are severely coupled. The research results of literature [1] show that the vibration shape of the symmetrical section beam is slightly affected by the speed. In this example, the speed changes mode shapes of the system obviously. The beams with unsymmetrical cross-sections show different characteristics compared with symmetric cross-sectional beams, for which effects of spinning speed on mode shapes are marginal. In addition, it can be seen from Fig. 3.3 that the rotation causes the mode shape to be a complex mode. This is due to the gyroscopic term in the system dynamics equation induced by the rotation, which can only be decoupled in the complex space. For each mode, the z-direction bending mode and the x-direction

Table 3.4 Example III natural frequency (Hz)

Spinning speed parameter	Natural frequency				
	#1	#2	#3	#4	#5
$\Omega^* = 0$	0.565	1.947	2.491	3.185	7.321
$\Omega^* = 0.2$	0.560	1.951	2.493	3.185	7.321
$\Omega^* = 0.4$	0.547	1.961	2.498	3.187	7.320
$\Omega^* = 0.6$	0.523	1.977	2.506	3.191	7.319
$\Omega^* = 0.8$	0.491	1.999	2.518	3.196	7.318
$\Omega^* = 1$	0.448	2.027	2.531	3.203	7.316

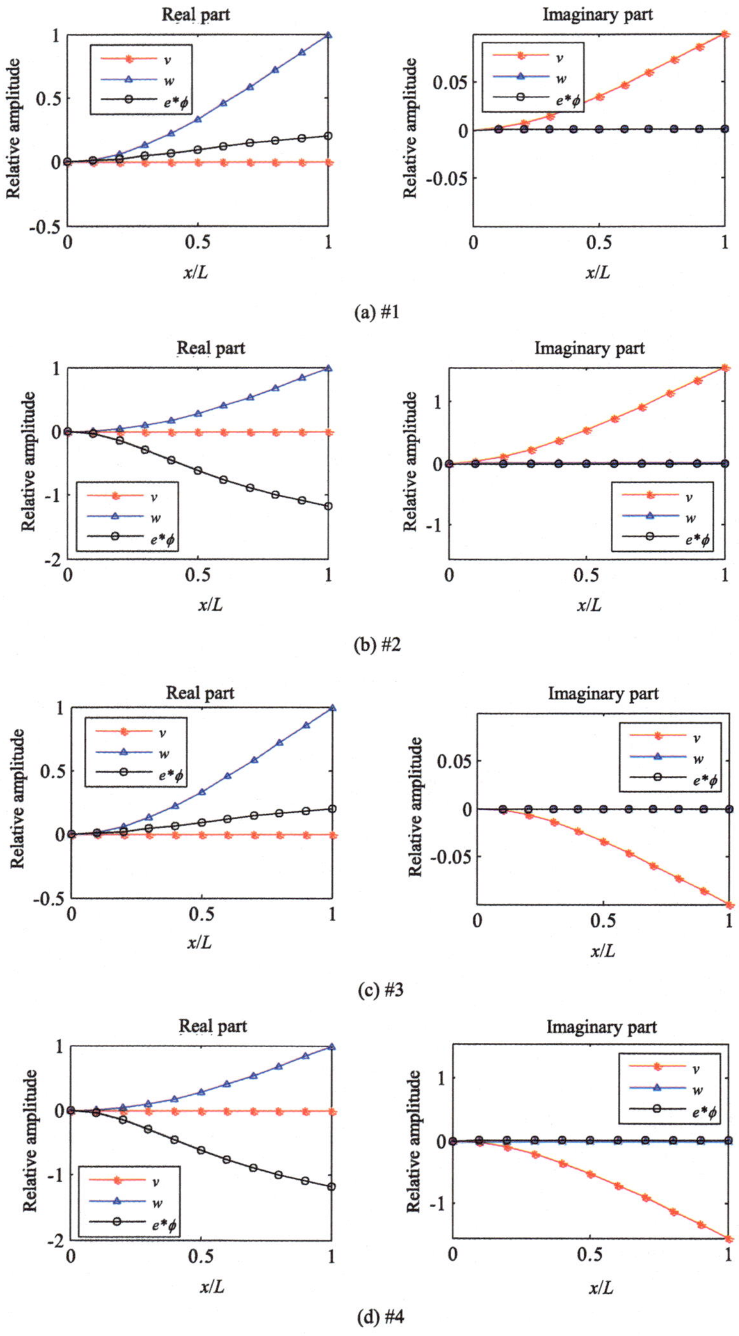

Fig. 3.3 Mode shapes of Example III with speed parameter for $\Omega^* = 1$

torsional mode have the same phase, and there is a 90° phase difference between the y-direction bending mode and the z-direction bending mode.

3.2.1.4 Effects of Spinning Speed

To examine the effect of the spinning speed on the natural frequencies of a beam with unsymmetrical cross section, various values with the interval [0, 4] for the spinning speed parameter are considered for the Example I and Example II, and corresponding frequencies are presented in Tables 3.5 and 3.6.

It is found that the spinning speed alters the natural frequencies, especially at the lower vibration modes. With the increase of the spinning speed, the coupling between y-axial and z-axial deformations becomes larger, which is demonstrated in Eq. (2.12). Therefore, mode shapes of the system change due to the larger coupling and natural frequencies vary correspondingly. Mostly, as the modal index rises, the effect of spinning speed on natural frequencies weakens since motion amplitudes become smaller with increasing frequency, which corresponds to an insignificant change in reference kinetic energy.

Table 3.5 Natural frequencies of Example I versus spinning speed parameter

Spinning speed parameter (Ω^*)	Natural frequency parameters (ω_i^*)					
	ω_1^*	ω_2^*	ω_3^*	ω_4^*	ω_5^*	ω_6^*
0	2.149	4.639	5.044	9.378	16.333	22.365
1	1.783	4.607	5.453	9.355	16.319	22.340
2	0.760	4.729	6.235	9.287	16.278	22.268
2.25	0	4.783	6.453	9.262	16.264	22.244
3	–	4.988	7.132	9.174	16.212	22.163
4	–	5.327	8.069	9.027	16.122	22.043

Table 3.6 Natural frequencies of Example II versus spinning speed parameter

Spinning speed parameter (Ω^*)	Natural frequency parameters (ω_i^*)					
	ω_1^*	ω_2^*	ω_3^*	ω_4^*	ω_5^*	ω_6^*
0	2.426	3.984	8.587	12.809	24.966	33.541
1	1.943	4.461	8.737	12.741	25.038	33.517
2	1.024	5.335	9.166	12.547	25.249	33.448
2.63	0	5.937	9.553	12.372	25.449	33.380
3	–	6.300	9.811	12.255	25.588	33.332
4	–	7.303	10.532	11.971	26.038	33.171

Figures 3.4 and 3.5 show variations of the first four non-dimensional natural frequencies with respect to the spinning speed parameter. Because of the large difference between the bending rigidities in the two principal planes natural frequencies start off with different values. The fundamental frequencies of both examples decrease with increasing spinning speed while the others decrease or increase. At a certain spinning speed, which is defined as the critical speed, the first natural frequency becomes negative, resulting in instability. For a spinning beam with circular or rectangular cross-section, the natural frequencies are obtained by subtracting or adding the natural frequencies when $\Omega^* = 0$ to the spinning speed parameter. So the value of the critical spinning speed when the beam becomes unstable equals to the first frequency of the beam with $\Omega^* = 0$. For a spinning beam with unsymmetrical cross-section, noncoincidence of mass center and shear

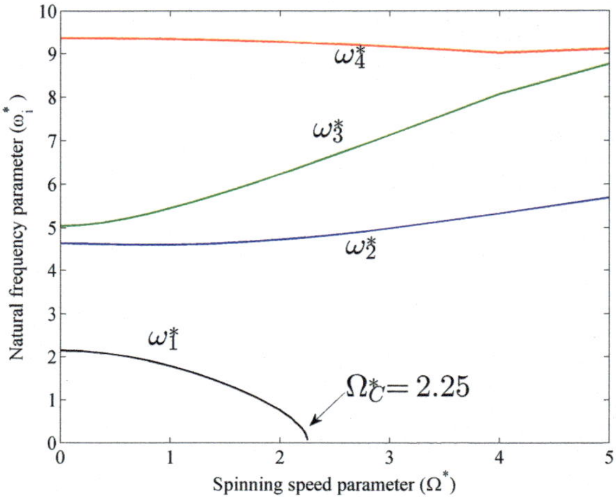

Fig. 3.4 Natural frequencies versus spinning speed for Example I

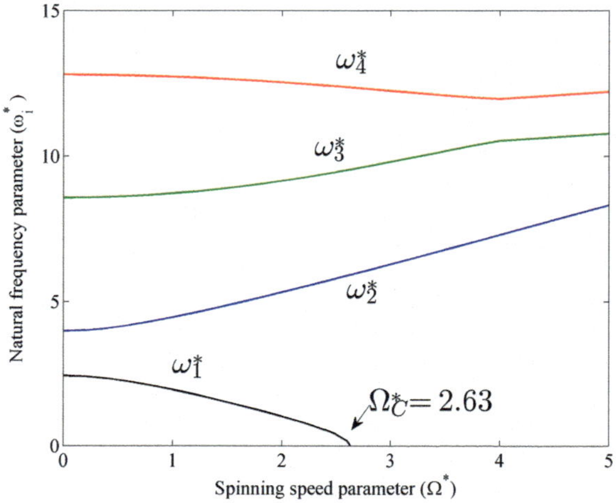

Fig. 3.5 Natural frequencies versus spinning speed for Example II

center induces coupled flexural–torsional modes and alters the critical speed. Both values of critical speed are larger than the first frequencies for examples studied.

3.2.1.5 Warping Effect

The relative errors of natural frequencies due to the warping effect are discussed in this section. Figures 3.6 and 3.7 show changes of natural frequencies with respect to spinning speed with inclusion and exclusion of the warping for Example I and II, respectively.

It is evident that the inclusion of the warping effect term increases the natural frequencies. And when the warping effect is neglected, the errors associated with

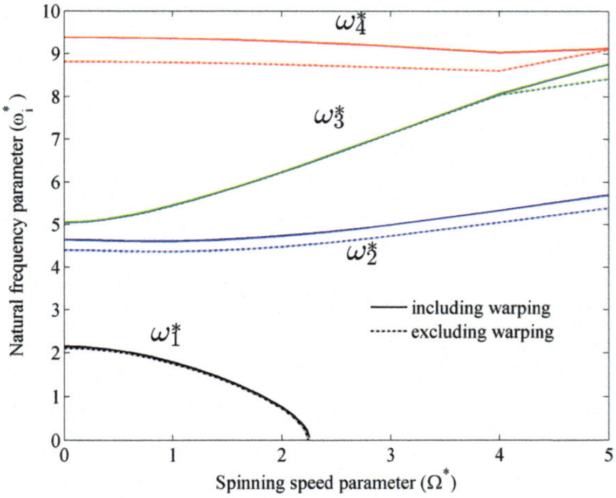

Fig. 3.6 Natural frequencies versus spinning speed for Example I

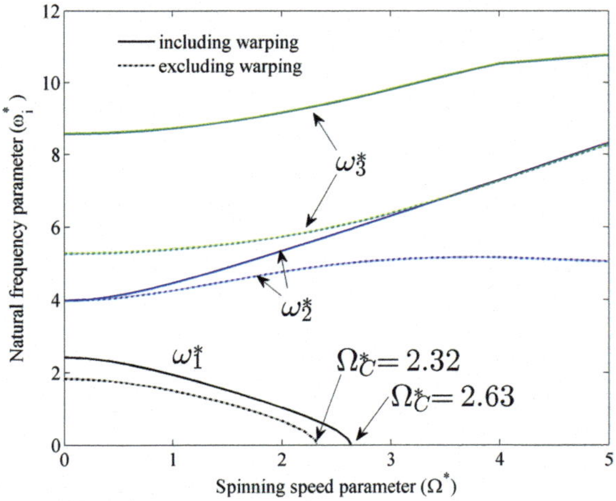

Fig. 3.7 Natural frequencies versus spinning speed for Example II

it become increasingly larger as the modal index increases. Additionally, errors of Example II are more severe than Example I. This is because the proportion of warping rigidity to bending rigidity of Example II is larger than that of Example I. It is also observed that exclusion of warping makes the values of critical speed decrease.

Through the above simulation, the following conclusions can be drawn:

(a) The dynamic model of flexible appendage with rigid body motions established in Chap. 2 can accurately describe the influence of rotation, eccentricity, warping and other factors on characteristics of the beam with an asymmetric cross-section;
(b) Unlike beams with symmetrical cross-sections, rotation along the axial direction changes the mode shape of the system, especially for low-order modes;
(c) After considering the warping effect, the critical speed of the system is increased. For systems with larger warping stiffness, the effect of increasing the critical speed is more significant. Moreover, effects of warping on natural frequencies become increasingly large when the proportion of warping rigidity to bending rigidity is notable.

3.2.2 Characteristics of Spinning Solar Wing

This section uses spinning beam-plate combined structure model of solar wing established in Sect. 2.3.3 to analyze the influence of the rotation rate on the dynamic characteristics. The solar wing is composed of a yoke and three solar panels. The root of the yoke is fixed, and the solar wing rotates at a constant angular velocity Ω around the x_s axis. The materials and physical properties of the yoke and solar panels are shown in Table 3.7. Regardless of the flexibility of the connecting hinge between the yoke and solar panel, the rigid connection is adopted for simulation. The finite element model is shown in Fig. 3.8.

Table 3.7 Physical properties of solar wing

Items	Yoke	Panel
Material	Aluminum alloy	Aluminum honeycomb sandwich panel
Young's modulus (GPa)	70	28
Poisson's ratio	0.3	0.3
Density (kg/m^3)	2700	150

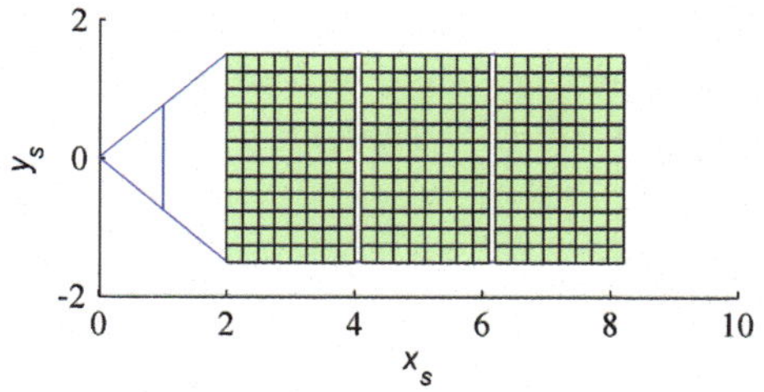

Fig. 3.8 Schematic diagram of solar wing finite element model

3.2.2.1 Natural Frequencies and Mode Shapes

Calculate the constrained mode of the cantilevered solar wing, that is, the root of the solar wing is fixed, as shown in Table 3.8. Figure 3.9 shows the first six mode shapes

Table 3.8 Description of mode shape of the solar wing

Mode	Frequency (Hz)	Mode description	Mode	Frequency (Hz)	Mode description
#1	0.318	1st out-of-plane bending	#6	5.185	2nd torsion
#2	0.857	1st in plane bending	#7	8.292	4th out-of-plane bending
#3	1.774	1st torsion	#8	8.508	3rd torsion
#4	1.797	2nd out-of-plane bending	#9	12.169	5th out-of-plane bending
#5	4.514	3rd out-of-plane bending	#10	12.486	6th out-of-plane bending

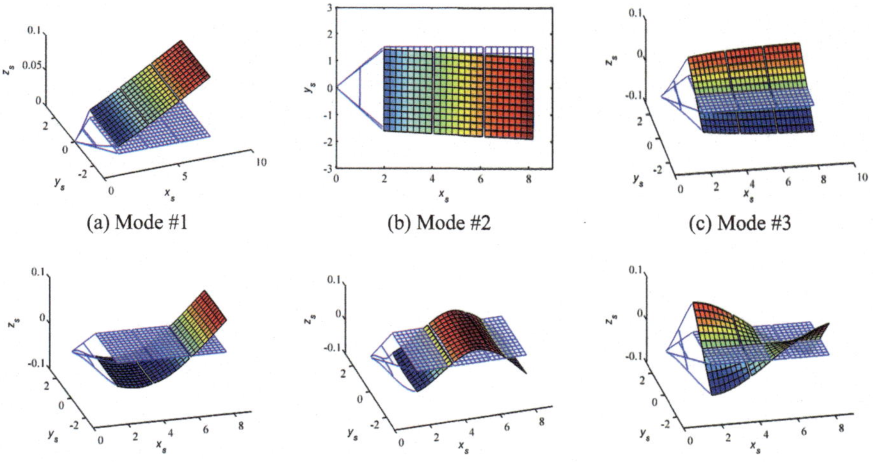

(a) Mode #1 (b) Mode #2 (c) Mode #3

(d) Mode #4 (e) Mode #5 (f) Mode #6

Fig. 3.9 Mode shapes of the cantilevered solar wing

of the solar wing. The fundamental frequency of the solar wing is 0.318 Hz, and the mode shape is the out-of-plane bending mode. The first ten modes can be divided into three vibration patterns: out-of-plane bending, in-plane bending and torsion.

3.2.2.2 Effects of Spinning Speed on Natural Frequencies

In order to investigate the influence of the spinning rate on the natural frequencies of the solar wing, the dimensionless speed and the dimensionless frequency are defined. The dimensionless speed is the ratio of the spinning speed to the fundamental frequency of the solar wing. The dimensionless frequency is the ratio of frequency when the solar wing is spinning to the frequency when the spinning speed is zero

$$\Omega^* = \Omega/\omega_{1,0} \quad \omega_i^* = \omega_i/\omega_{i,0} \tag{3.3}$$

In the formula, $\omega_{1,0}$ is the fundamental frequency of the solar wing when the spinning speed is zero, and $\omega_{i,0}$ is the i-th order frequency of the solar wing when the spinning speed is zero.

Figures 3.10, 3.11 and 3.12 present the influence curves of the dimensionless speed on the first three natural frequencies. It can be seen that with the increase of speed, the fundamental frequency of the solar wing shows a decreasing trend. When the dimensionless speed is increased from 0 to 0.1 (speed is 11.48 °/s), the fundamental frequency is reduced from 0.318 to 0.315 Hz (0.8% reduction), and when it is further increased to 0.5, the fundamental frequency is reduced to 0.257 Hz (19.2% reduction). The second-order frequency shows an increasing trend, which is consistent with the results of the analysis using the flexible beam model.

Figures 3.13, 3.14 and 3.15 show the influence of the dimensionless speed on the dimensionless natural frequency of the out-of-plane bending modes, the in-plane modes and the torsion modes respectively. It can be seen that the speed has a greater influence on the low-order natural frequencies.

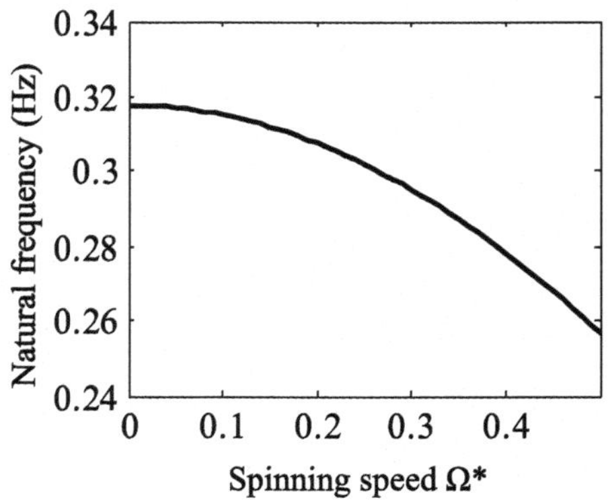

Fig. 3.10 The influence of speed on the first natural frequency

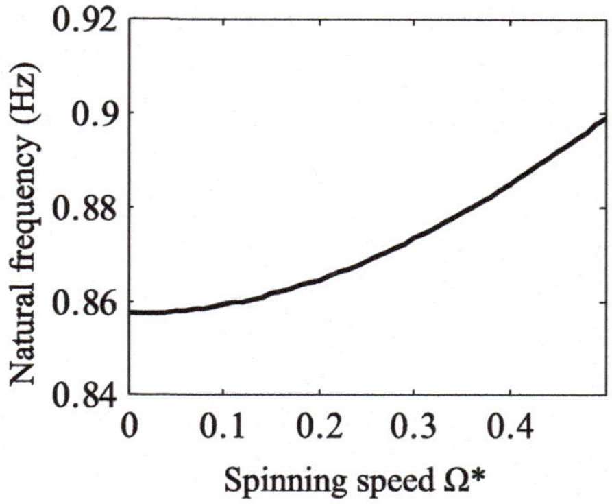

Fig. 3.11 The influence of speed on the second natural frequency

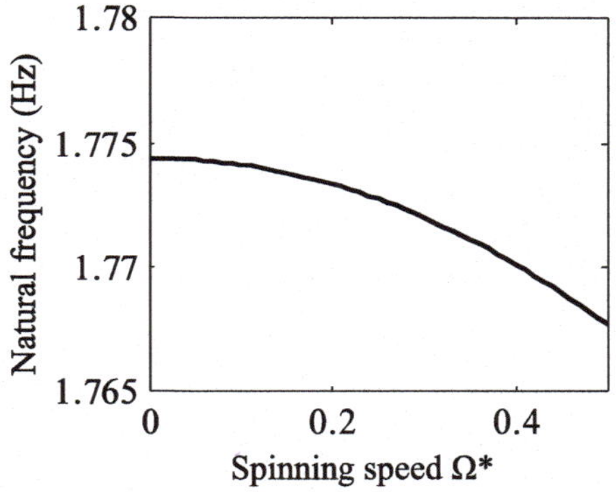

Fig. 3.12 The influence of speed on the third natural frequency

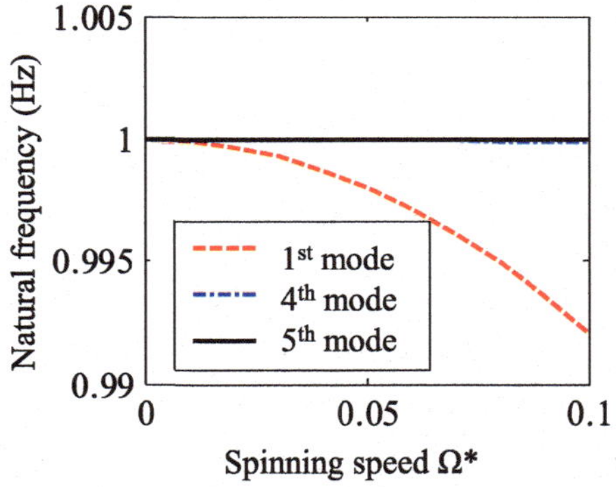

Fig. 3.13 The influence of speed on the natural frequency of out-of-plane bending modes

3.3 Dynamic Analysis of Flexible Spacecraft

Based on the dynamic model of the flexible spacecraft established in Sect. 2.4, this section analyzes the influence of the rotation of the flexible appendage on the system

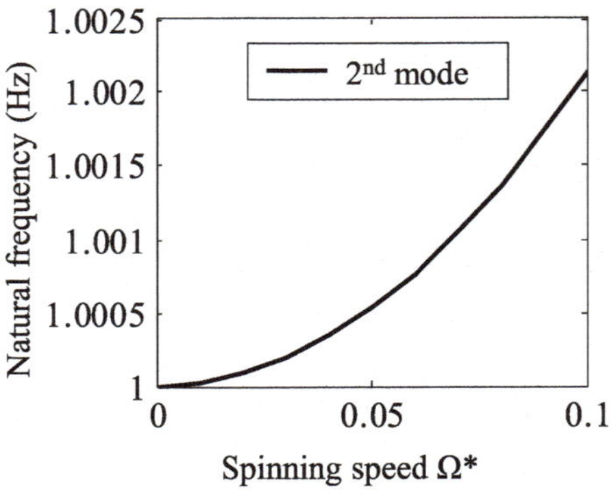

Fig. 3.14 The influence of speed on the natural frequency of in-plane bending mode

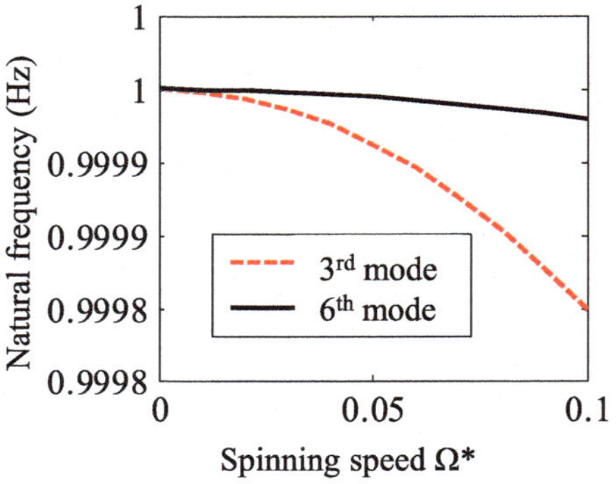

Fig. 3.15 The influence of speed on the natural frequency of torsion modes

parameters and the coupling between the flexible appendage and the spacecraft platform.

3.3.1 Simulation Case

In order to evaluate the influence of the rotation of the flexible appendage, a numerical simulation example is given in this section. Figure 3.16 shows the model of a typical spacecraft, which consists of a platform and two symmetrical flexible solar wings. The solar wing consists of a yoke and three solar panels. The solar wings are hinged on the surface of the platform and can rotate about the x_s axis at a constant angular velocity Ω, which is parallel to the x axis. Let the symbol γ denotes the angle between the body frames $oxyz$ and $o_s x_s y_s z_s$. The position vector from point o to o_s is $\boldsymbol{r}_{os} = [1\ \text{m}, 0, 0]^{\text{T}}$. The total mass of the solar wing is 118.6 kg. The inertia matrix of the platform along axes xyz is given by $\boldsymbol{J} = diag\ [2666.7, 2666.7, 1666.7]$ kg·m^2 and the inertia matrix of the solar wing along axes $x_s y_s z_s$ is given by $\boldsymbol{J} = diag\ [53.6, 1607.9, 1661.6]$ kg·m^2. The structural damping associated with the solar arrays is assumed to be zero.

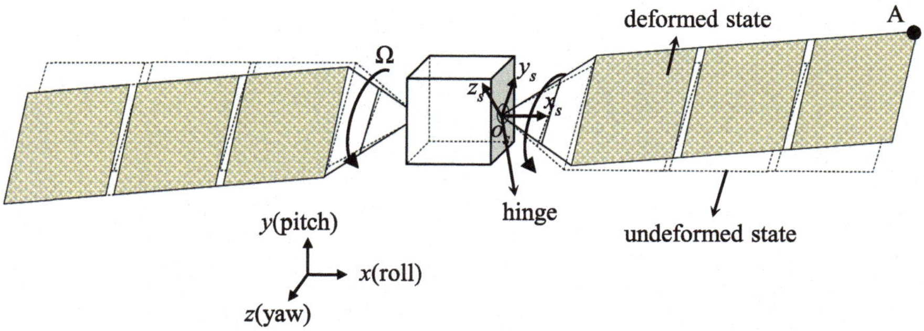

Fig. 3.16 Schematic of the spacecraft with rotating solar wings

In the dynamic model, the solar wing is idealized as the assembly of beams and plates. In physical systems, the solar array is a complex structure of the yoke, honeycomb and hinges. Simplifications in the ideal model would induce small differences in the results.

3.3.2 Influence of Rotation on System Parameters

3.3.2.1 Influence of Rotation on Inertia Matrix

When the solar wing is located at three positions relative to the platform ($\gamma = 0°$, 45 and 90°), the moment of inertia matrix is

$$
\begin{aligned}
\gamma = 0^o: \quad \boldsymbol{J} &= \begin{bmatrix} 2774 & & \\ & 7187.7 & \\ & & 5695 \end{bmatrix} \text{kg} \cdot \text{m}^2 \\
\gamma = 45^o: \quad \boldsymbol{J} &= \begin{bmatrix} 2774 & & \\ & 7241.4 & -53.7 \\ & -53.7 & 5641.4 \end{bmatrix} \text{kg} \cdot \text{m}^2 \\
\gamma = 90^o: \quad \boldsymbol{J} &= \begin{bmatrix} 2774 & & \\ & 7295 & 0 \\ & 0 & 5587.7 \end{bmatrix} \text{kg} \cdot \text{m}^2
\end{aligned}
\tag{3.4}
$$

It can be seen that the rotation of the solar wing does not change the inertia of the system around the x axis. Figure 3.17 shows the influence on the system moment of inertia matrix during the rotation of the solar wing. When γ changes from 0 to 90°, the spacecraft's moment of inertia around the y and z axis changes by 1.49 and 1.88%, respectively. In this simulation case, the difference between the moment of inertia around the axis y_s and around the z_s axis is small. When the difference is larger, the

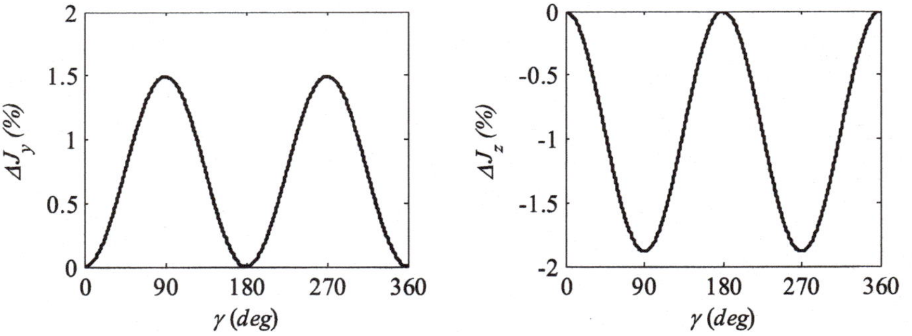

(a) Moment of inertia about the y axis (b) Moment of inertia about the z axis

Fig. 3.17 The influence of rotation on moment of inertia of spacecraft

influence of rotation on the moment of inertia will be more severe. When γ equals to 45°, the moment of inertia matrix is a non-diagonal matrix, and the rotation of the spacecraft around the y and z axes is coupled.

3.3.2.2 Influence of Rotation on the Rigid-Flex Coupling Matrix

For simplicity, we only consider the first six cantilevered flexible modes. The rigid-flex coupling matrix of the spacecraft is

$$\boldsymbol{D} = \begin{bmatrix} 0 & 0 & 8.433 & 0 & 0 & -3.4535 \\ 63.5\cos\gamma & -67.3\sin\gamma & 0 & 19.26\cos\gamma & 8.99\cos\gamma & 0 \\ 63.5\sin\gamma & 67.3\cos\gamma & 0 & 19.26\sin\gamma & 8.99\sin\gamma & 0 \end{bmatrix} \tag{3.5}$$

It can be noted that ω_x is only influenced by the third and sixth flexible modes, i.e., the x_s-axial torsional modes. Figure 3.18 shows coupling coefficients of the

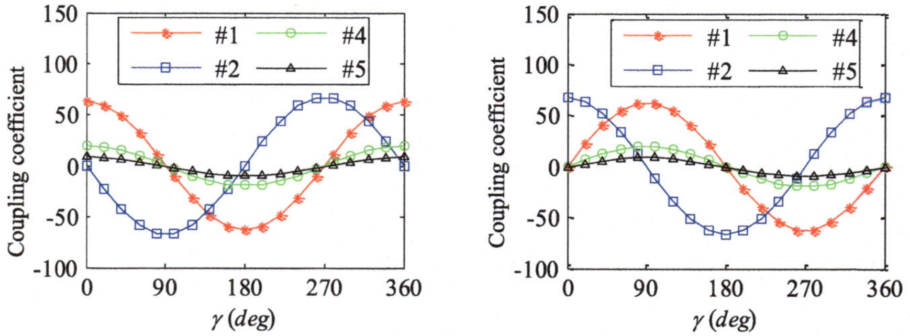

(a) y (b) z

Fig. 3.18 Coupling coefficient of modal coordinates during rotation of solar wing first three out-of-plane bending modes along axis y (corresponding to the #1, #4, #5 modal coordinates) and the first in-plane bending mode along axis z (corresponding to the #2 modal coordinate) with respect to the rotation angle. When the rotation angle γ equals to 0 or 180°, the y-direction attitude angular velocity ω_y and the z-direction attitude angular velocity ω_z are only coupled with the y_s-axial bending and z_s-axial bending modes, respectively. When the rotation angle γ equals to 90 or 270°, the y-direction angular velocity ω_y and the z-direction angular velocity ω_z are only coupled with the z_s-axial bending and y_s-axial bending modes, respectively. In other positions, the angular velocity ω_y, ω_z and bending modes along z_s and y_s axes are coupled with each other.

3.3.3 *Influence of Rotation on System Characteristics*

Natural frequencies of the system for different angular orientation of the solar wing are determined by linearizing the equations. And the results are compared to the ones given by commercial FEM software. The spacecraft's frequencies of the first six elastic modes are calculated with three different angular orientation of the solar wing: 0, 45 and 90°, as listed in Table 3.9. And mode shapes for $\gamma = 0°$ are shown in Fig. 3.19.

Frequencies calculated by the present method agree well with the results from FEM software. The minute differences between the present method and the commercial software are brought on by the different scales of the meshes. The more sophisticated mesh is applied in the software. Because of the large rigidity of the platform, deformations of the solar wing dominate mode shapes of the whole spacecraft. Within the first six elastic modes, there exist three bending modes along y_s axis, two torsional modes along x_s axis and one bending mode along z_s axis.

The frequency values for the first bending modes along y_s axis and z_s axis vary remarkably with different orientations of the solar wing. y_s bending modes are related to the inertia of the platform along the y_s axis while z_s bending modes related to the inertia along the y_s axis. As the solar wing rotates from 0 to 90°, the platform's inertia

Table 3.9 Calculated frequencies (Hz) of the satellite: (1) present method; (2) FEM

Mode index	0°		45°		90°		Description
	[1]	[2]	[1]	[2]	[1]	[2]	
#1	0.475	0.482	0.520	0.519	0.587	0.593	1st y_s bending
#2	1.894	1.922	1.609	1.631	1.392	1.410	1st x_s torsion
#3	1.798	1.803	1.798	1.803	1.798	1.803	1st z_s bending
#4	1.915	1.954	2.037	2.080	2.066	2.101	2nd y_s bending
#5	4.582	4.658	4.632	4.566	4.680	4.765	3rd y_s bending
#6	5.198	5.211	5.200	5.211	5.201	5.211	2nd x_s torsion

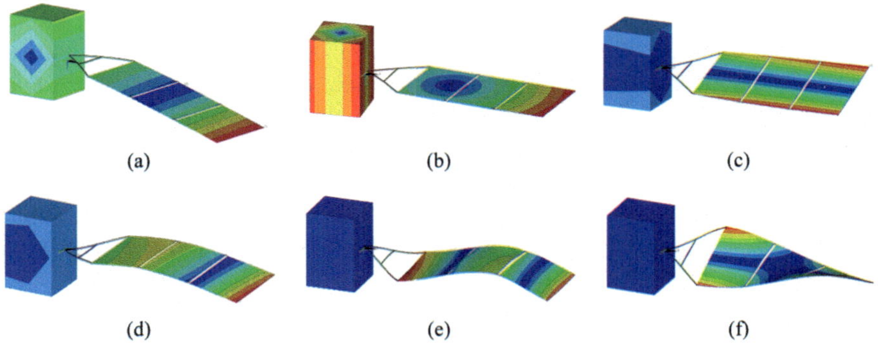

Fig. 3.19 Mode shapes of the spacecraft for $\gamma = 0°$: **(a)** Mode #1; **(b)** Mode #2; **(c)** Mode #3; **(d)** Mode #4; **(e)** Mode #5; **(f)** Mode #6;

along the y_s axis deceases so that the frequencies for y_s bending modes increase, which is especially prominent for the first mode. Similarly, the platform's inertia along the z_s axis increases so that the frequency for z_s bending mode decreases. The influence of the inertia weakens with the index of mode increases. Frequencies for the two torsional modes along x_s axis remain constant with respect to the orientation of the solar wing. This is due to the fact that these two modes are related to the inertia of the platform along the x_s axis, which remains constant when the solar wing rotates.

Figure 3.20 shows the singular-value Bode diagram of the open-loop system for the spacecraft when $\gamma = 0, 45$ and $90°$. The input is controlled torques of three-axes and the output is the attitude angles of the spacecraft. By comparison, when the solar wing rotates from 0 to 90°, the first-order frequency shifts to the right (increased), the second-order frequency shifts to the left (decreases), and the third-order frequency position remains unchanged.

3.3.4 Influence of Elastic Vibration on Attitude

In this experiment, transient response analyses of the system are conducted. To demonstrate the influence of the first order terms of the transformation matrix on the system dynamics, zero-input responses simulations are performed. Simulations are carried out in two cases as follows: (1) with the first order terms considered, i.e., the first order model and (2) with the first order terms neglected, i.e., the zero order model. In both cases, we give the same initial conditions that both Euler angles and angular velocities of the platform are zeros and the quasi-static deformation of the free end of the solar panel is assumed to be 0.1 m. Dynamical responses of the system are calculated in 30 s.

Figures 3.21, 3.22 and 3.23 show transient responses of the spacecraft with constant angular velocity $\Omega = 5$ °/s. Displacements of the point A along axes y_s and z_s are curved in Fig. 3.21. The maximum value of the elastic deflection of the

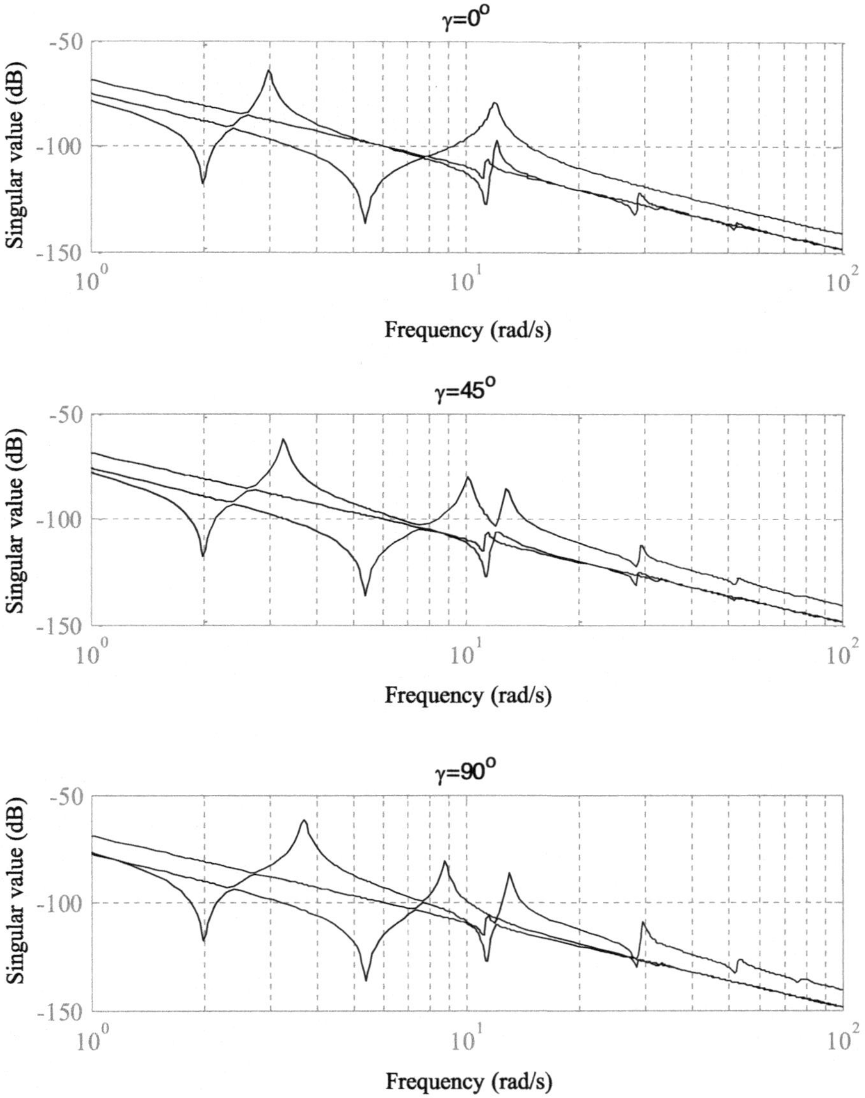

Fig. 3.20 Singular-value Bode diagram of open loop system for spacecraft

solar panel remains 0.1 m approximately. In Fig. 3.22, the angular velocities are reported. It is observed that no significant difference takes place when the first order terms is considered. Time histories of Euler's angles, evaluated with respect to the inertial reference frame, are reported in Figure 3.24 is the Euler angle deviation curve. This maximum difference is of the order of magnitude of 2 ° on the yaw angle.

Figures 3.25 and 3.28 present zero-input responses of the spacecraft with constant angular velocity $\Omega = 20$ °/s. Compared with results of $\Omega = 5$ °/s, the difference

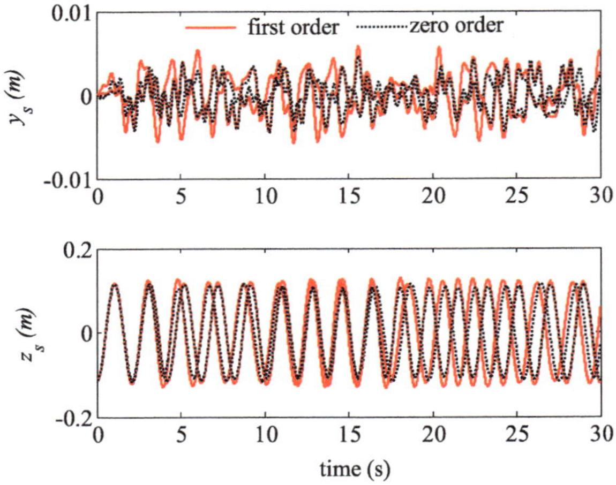

Fig. 3.21 Time histories of tip's displacements ($\Omega =$ 5 °/s)

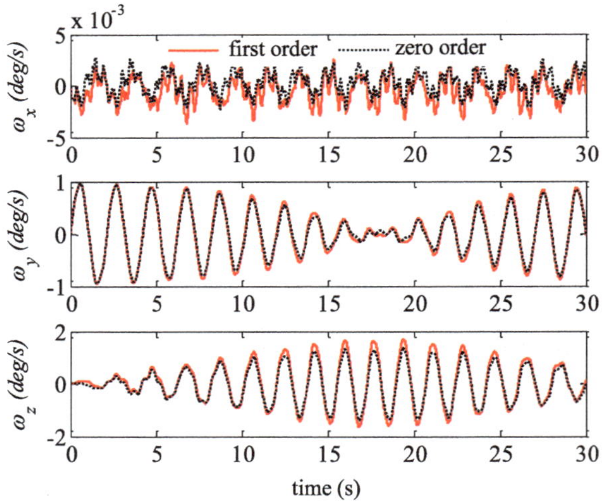

Fig. 3.22 Time histories of angular velocities ($\Omega = 5$ °/s)

between two models increases with the increase of angular velocity. As the rotating velocity is increased, it can be noted that, since the variation of the transformation matrix with respect to time increases, the first order model should be adapted for the purpose of accurately predicting the system dynamics.

During the transient analysis, the moments of inertia matrix of the spacecraft is assumed to remain constant with respect to time. In physical systems, the moments of inertia matrix are time-varying due to the consumption of fuel. Moreover, the elastic deformation of the flexible appendage would slightly alter the moments of inertia of the system. Nevertheless, the results reveal the several errors of the zero-order model (Figs 3.25, 3.26, 3.27 and 3.28).

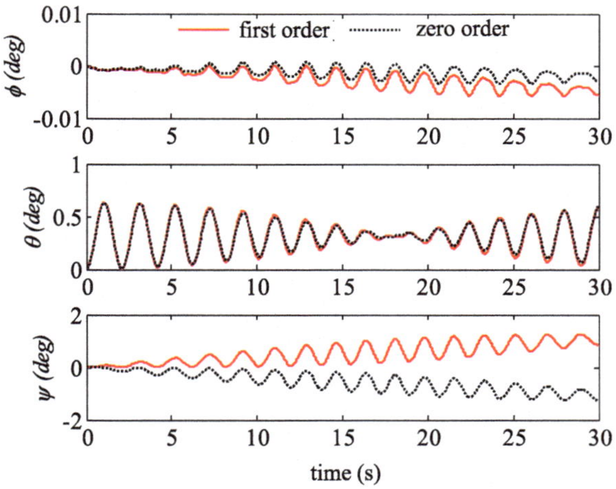

Fig. 3.23 Time histories of Euler angles ($\Omega = 5$ °/s)

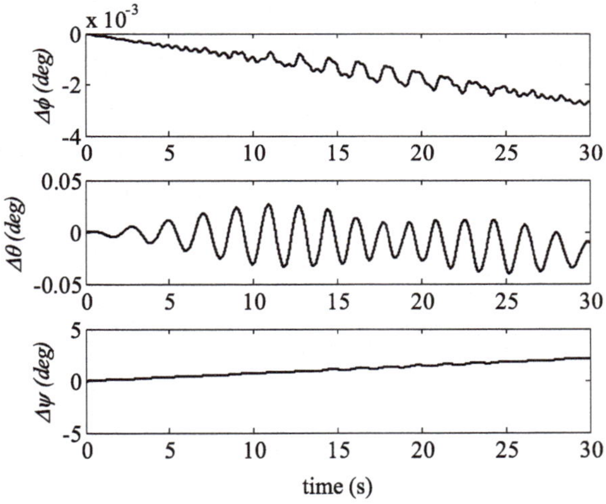

Fig. 3.24 Euler angle deviation ($\Omega = 5$ °/s)

3.4 Summary

Based on the models established in the previous chapter, this chapter carried out the dynamic analysis of the flexible spacecraft with time-varying parameters. The conclusions obtained are as follows:

(1) For a rotating flexible beam, the system control equation can only be discretized by complex modes because of the gyro terms in the system governing equation. Also, both the axial rotation and warping changes natural frequencies and mode shapes of the system. And it has a greater impact on low-order modes.
(2) The constrained modes and natural frequencies of a solar wing have been obtained, and the influence of the rotation speed on the natural frequencies

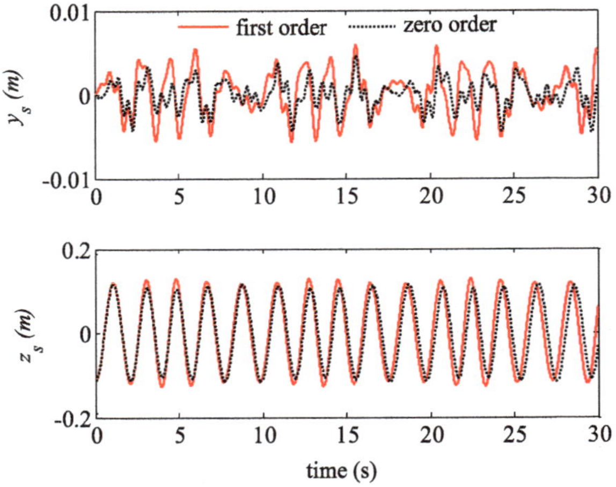

Fig. 3.25 Time histories of tip's displacements ($\Omega = 20$ °/s)

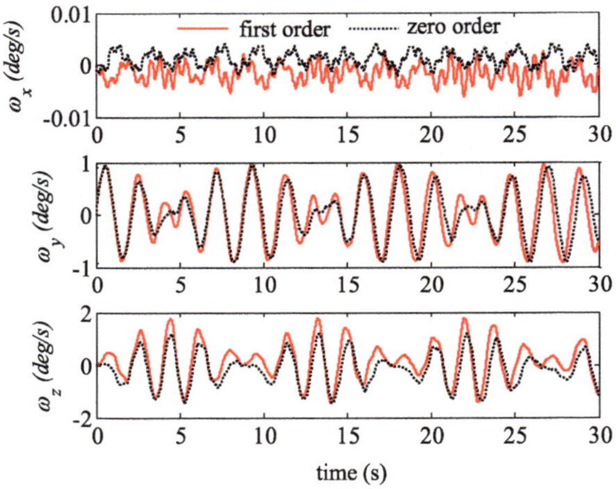

Fig. 3.26 Time histories of angular velocities ($\Omega = 20$ °/s)

of the solar wing has been analyzed. The results show that when the speed increases, the natural frequencies of each order show a decreasing trend, especially for the lower order.

(3) Transient response analyses of a central platform with rotating solar wings have been conducted in order to evaluate the influence of the first-order terms. Numerical results have shown that the effects of first-order terms could be significant during the transient vibration motion of the flexible appendages. When the appendage rotates with a growing angular velocity, severe errors occurred in the responses of the conventional model.

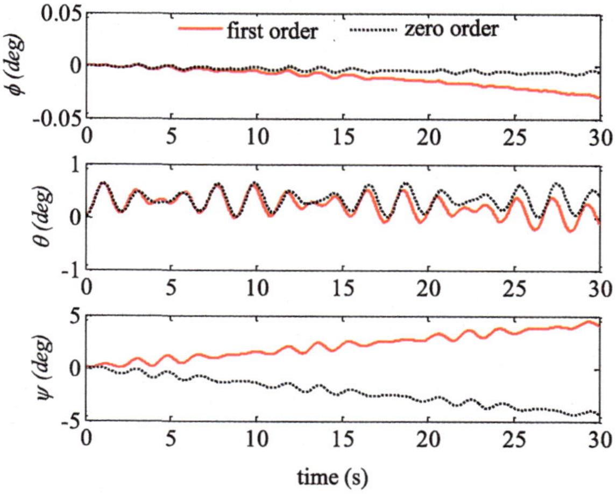

Fig. 3.27 Time histories of Euler angles ($\Omega = 20$ °/s)

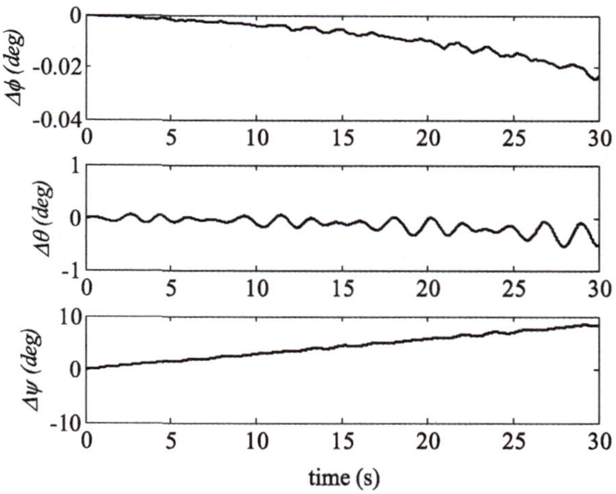

Fig. 3.28 Euler angle deviation ($\Omega = 20$ °/s)

References

1. Banerjee, J. R., Su, H. (2004). Development of a dynamic stiffness matrix for free vibration analysis of spinning beams. *Computers and Structures, 82*, 2189–2197
2. Bercin, A. N., & Tanaka, M. (1997). Coupled flexural-torsional vibrations of timoshenko beams. *Journal of Sound and Vibration, 207*(1), 47–59.

Chapter 4
Vibration Control Methods for Systems in Complex Mode Space

4.1 Introduction

In the coupling dynamics analysis of the system in the previous chapter, it can be seen that the elastic vibration of the flexible appendage is one of the main disturbance sources affecting the attitude of the spacecraft. Therefore, the vibration control of flexible appendage is significant in improving the pointing accuracy and attitude of the spacecraft. The main purpose of this chapter is to study the vibration control method of the flexible appendage, and design effective controllers for the flexible appendage to attenuate elastic vibrations. In the complex space, the design methods of state feedback controller, sliding mode controller and positive position feedback controller are proposed, and numerical simulations are carried out to verify the effectiveness of the control methods.

For last decades, numerous studies have been conducted by researchers and various control techniques have been proposed for different mission objectives, such as optimum control [1] , sliding mode control (SMC) [2–6], independent modal space control [7, 8], positive position feedback [9], robust control method [10], adaptive control method [11], and fuzzy control method, etc. However, according to previous literature, most systems studied belong to the real space, i.e., the state-space matrices of the systems remained to be real number, which is not the case when the structural damping and gyroscope effects are considered. Few papers were concentrated on the vibration control of complex mode systems. Shao and Zhang [12] developed a hybrid independent modal controller on the basis of complex mode theory for high-speed flexible linkage mechanisms. The controller employed the state feedback law to allocate the poles to provide sufficient damping for the closed-loop system. They also designed robust H_∞ controller, which was essentially developed in real number space [13]. It is necessary to study the active vibration control method in complex space for the flexible structure of the spacecraft.

© The Author(s), under exclusive license to Springer Nature Singapore Pte Ltd. 2022

J. Wang and D.-X. Li, *Rigid-Flexible Coupling Dynamics and Control of Flexible Spacecraft with Time-Varying Parameters*,
https://doi.org/10.1007/978-981-16-5097-0_4

4.2 Linear State Feedback Stabilization in Complex Mode Space[1]

This section proposes the design method of the state feedback controller for the complex modal system. The design of the control system is based on Lyapunov's theory. In general, the system is mainly affected by low-order modes, so only the first $2n$-order modes are retained, as presented in Eq. (2.86). The governing equation of the system is rewritten as

$$\begin{aligned} \dot{\boldsymbol{x}} &= \boldsymbol{A}\boldsymbol{x} + \boldsymbol{B}\boldsymbol{u} + \boldsymbol{d} \\ \boldsymbol{V}_s &= \boldsymbol{C}\boldsymbol{x} \end{aligned} \tag{4.1}$$

Assuming enough sensors are distributed on the beam so that all the states in Eq. (4.1) are detectable. Ignoring the external load, then the system can be expressed as

$$\dot{\boldsymbol{x}} = \boldsymbol{A}\boldsymbol{x} + \boldsymbol{B}\boldsymbol{u} \tag{4.2}$$

The design method of the state feedback controller and the gain scheduling controller when the input saturation is considered is given below.

4.2.1 State Feedback Controller

Lemma 4.1 *Consider hermitian solutions of the algebraic Riccati equation of the form* [14].

$$\boldsymbol{P}^H\boldsymbol{A} + \boldsymbol{A}^H\boldsymbol{P} - \boldsymbol{P}\boldsymbol{B}\boldsymbol{B}^H\boldsymbol{P} + \boldsymbol{Q} = 0 \tag{4.3}$$

where $\boldsymbol{A}$, $\boldsymbol{B}$, $\boldsymbol{Q}$ are complex matrices of dimensions of $n \times n$, $n \times m$, and $n \times n$, respectively, $\boldsymbol{Q}$ is positive semi-definite, or is just hermitian. Further the pair ($\boldsymbol{A}$, $\boldsymbol{B}$) is assumed to be stabilizable. This means that there is an $m \times n$ matrix $\boldsymbol{K} = -\boldsymbol{B}^{\mathrm{H}}\boldsymbol{P}$ such that $\boldsymbol{A}$-$\boldsymbol{BK}$ is asymptotically stable (i.e., all the eigenvalues of $\boldsymbol{A}$-$\boldsymbol{BK}$ lie in the open left half plane).

Theorem 4.1 *Considers system in Eq. (4.2). Assume that ($\boldsymbol{A}$, $\boldsymbol{B}$) is stabilizable, there exists a positive definite solution $\boldsymbol{P}(2n \times 2n)$ for any positive definite matrix $\boldsymbol{Q}$such that*

[1] J. Wang, J. Wu and G. Wu, "Time-varying feedback stabilization of systems with input saturation actuators in complex mode space," 2020 Chinese Automation Congress (CAC), 2020, pp. 3074–3079.

$$P^H A + A^H P - PBB^H P + Q = 0 \tag{4.4}$$

Moreover, this positive definite solution P has the following properties:

(i) Partition the positive definite matrix Q and solution P into square sub-matrices as follows

$$Q = \begin{bmatrix} Q_{11} & Q_{12} \\ Q_{21} & Q_{22} \end{bmatrix} \qquad P = \begin{bmatrix} P_{11} & P_{12} \\ P_{21} & P_{22} \end{bmatrix} \tag{4.5}$$

Q_{ij} and $P_{ij}(i,j = 1, 2)$ are all matrices of dimensions $n \times n$. If Q_{ij} subject to conditions as follows

$$Q_{22} = Q_{11}^* \qquad Q_{12}^T = Q_{12} \tag{4.6}$$

Then the sub-matrices satisfy

$$P_{22} = P_{11}^* \qquad P_{12}^T = P_{12} \tag{4.7}$$

(ii) For the solution matrix P subject to Eq. (4.7), the state feedback controller of the form

$$u = -B^H Px \tag{4.8}$$

satisfy $u^* = u$ can make the system in Eq. (4.2) stable.

(iii) The eigenvalues of A are all locate in the closed left-half plane.

Proof From Lemma 4.1, it is obvious that the solution P exists.

(i) The conjugate system for system given in Eq. (4.2) can be expressed as

$$\dot{x}^* = A^* x^* + B^* u^* \tag{4.9}$$

Note that the control input u is a real vector (i.e., $u \in \mathbb{R}^m$, $u^* = u$) for a physical system. Referring to Eq. (4.4), the Riccati equation can be written as

$$P^T A^* + A^T P^* - P^* B^* B^T P^* + Q^* = 0 \tag{4.10}$$

Let us define a transformation matrix $\boldsymbol{T}$ as follows

$$\boldsymbol{T} = \begin{bmatrix} 0 & \boldsymbol{I}_{n\times n} \\ \boldsymbol{I}_{n\times n} & 0 \end{bmatrix} \quad s.t. \quad \boldsymbol{T} = \boldsymbol{T}^{-1} = \boldsymbol{T}^T \tag{4.11}$$

According to Eq. (2.87), components in Eq. (4.9) satisfy the following form

$$\boldsymbol{x}^* = \boldsymbol{T}\boldsymbol{x} \qquad \boldsymbol{A}^* = \boldsymbol{T}^{-1}\boldsymbol{A}\boldsymbol{T} \qquad \boldsymbol{B}^* = \boldsymbol{T}^{-1}\boldsymbol{B} \tag{4.12}$$

Based on such transformation, the Eq. (4.10) can be cast into the following form

$$(\boldsymbol{T}\boldsymbol{P}^*\boldsymbol{T})^H\boldsymbol{A} + \boldsymbol{A}^H(\boldsymbol{T}\boldsymbol{P}^*\boldsymbol{T}) - (\boldsymbol{T}\boldsymbol{P}^*\boldsymbol{T})\boldsymbol{B}\boldsymbol{B}^H(\boldsymbol{T}\boldsymbol{P}^*\boldsymbol{T}) + \boldsymbol{T}\boldsymbol{Q}^*\boldsymbol{T} = 0 \tag{4.13}$$

Based on Eq. (4.6), there follows

$$\boldsymbol{T}\boldsymbol{Q}^*\boldsymbol{T} = \begin{bmatrix} \boldsymbol{Q}_{22}^* & \boldsymbol{Q}_{21}^* \\ \boldsymbol{Q}_{12}^* & \boldsymbol{Q}_{11}^* \end{bmatrix} = \begin{bmatrix} \boldsymbol{Q}_{11} & \boldsymbol{Q}_{12} \\ \boldsymbol{Q}_{21} & \boldsymbol{Q}_{22} \end{bmatrix} = \boldsymbol{Q} \tag{4.14}$$

Using Eqs. (4.13) and (4.4), one obtains

$$\boldsymbol{T}\boldsymbol{P}^*\boldsymbol{T} = \boldsymbol{P} \Rightarrow \begin{bmatrix} \boldsymbol{P}_{22}^* & \boldsymbol{P}_{21}^* \\ \boldsymbol{P}_{12}^* & \boldsymbol{P}_{11}^* \end{bmatrix} = \begin{bmatrix} \boldsymbol{P}_{11} & \boldsymbol{P}_{12} \\ \boldsymbol{P}_{21} & \boldsymbol{P}_{22} \end{bmatrix} \Rightarrow \begin{cases} \boldsymbol{P}_{22} = \boldsymbol{P}_{11}^* \\ \boldsymbol{P}_{12}^T = \boldsymbol{P}_{12} \end{cases} \tag{4.15}$$

Hence, the Eq. (4.7) is demonstrated.

(ii) According to Lemma 4.1, the state feedback controller given in Eq. (4.8) is such that $\boldsymbol{A}$-$\boldsymbol{B}\boldsymbol{B}^H\boldsymbol{P}$ is Hurwitz-stable.

Moreover, the conjugate of the input is

$$\boldsymbol{u}^* = -(\boldsymbol{B}^*)^H\boldsymbol{P}^*\boldsymbol{x}^* \tag{4.16}$$

Inserting Eq. (4.7) into Eq. (4.12), leading to

$$\boldsymbol{u}^* = -(\boldsymbol{T}^{-1}\boldsymbol{B})^H\boldsymbol{T}\boldsymbol{P}^*\boldsymbol{T}\boldsymbol{T}\boldsymbol{x} = -\boldsymbol{B}^H\boldsymbol{P}\boldsymbol{x} = \boldsymbol{u} \tag{4.17}$$

Therefore the input $\boldsymbol{u}$ is a real vector and meets the actual requirements of the physical system.

(iii) As listed in Eq. (2.51), sub-matrices $\boldsymbol{M}$ and $\boldsymbol{K}$ are both positive-definite and the matrix $\boldsymbol{G}$ is anti-symmetric. So eigenvalues of $\boldsymbol{A}$ are all locate on the imaginary axis of the complex plane.

The proof is finished.

Theorem 4.1 provides a form of the state feedback controller for systems with complex modes. Notice that the controller is derived by extending the conventional linear quadratic regulator theory. Through this method one can get reasonably controllers that can achieve the expected performance. Nevertheless, the design of the controller is not restricted by the saturation of the input, which is frequently encountered in actual systems. In the sequel, we develop a state feedback law with the saturation considered.

4.2.2 Gain Scheduled Controller

In most of the practical systems, the actuator capacity is limited. For the system studied in the present paper, the input voltages of the piezoelectric patches, denoted by $\boldsymbol{u}$, should respect to its boundaries. In this section, we consider the control design under actuator constraints.

For the amplitude limitation of actuator $\boldsymbol{u}$, a nonlinear dynamic model is used in the general form

$$\dot{\boldsymbol{x}} = \boldsymbol{A}\boldsymbol{x} + \boldsymbol{B}sat(\boldsymbol{u}) \tag{4.18}$$

where matrices $\boldsymbol{A}$ and $\boldsymbol{B}$ are given in Eq. (4.2). Function $sat(\boldsymbol{u})$, which represents actuator saturation and is an $m \times 1$ vector with each element $sat(\boldsymbol{u}_j)$ defined as

$$sat(\boldsymbol{u}_j) = sign(\boldsymbol{u}_j) \times Min\{\boldsymbol{u}_j^{\max}, \|\boldsymbol{u}_j\|\} \tag{4.19}$$

where $\boldsymbol{u}_j^{\max}$ is the maximum value of the j-th input and sign(•) denotes the sign function.

First, we introduce the design of low gain law, then the method which combines the gain scheduled control and nested control is developed.

Lemma 4.2 *Let* $\boldsymbol{Q}_\varepsilon: (0,1] \to \mathbb{C}^{n\times n}$ *be any matrix that is positive definite for any* $\varepsilon \in (0,1]$ *and satisfies*

$$\frac{d\boldsymbol{Q}_\varepsilon}{d\varepsilon} > 0 \quad and \quad \lim_{\varepsilon\to 0} \boldsymbol{Q}_\varepsilon = 0 \tag{4.20}$$

$\boldsymbol{X} \geq \boldsymbol{Y}$ for hermitian matrices $\boldsymbol{X}$ and $\boldsymbol{Y}$ means that $\boldsymbol{X}$–$\boldsymbol{Y}$ is positive semi-definite.

For each $\varepsilon \in (0,1]$, there exists a unique positive definite solution $\boldsymbol{P}_\varepsilon$ that solves the algebraic Riccati equation defined as

$$\boldsymbol{P}^H\boldsymbol{A} + \boldsymbol{A}^H\boldsymbol{P} - \boldsymbol{P}\boldsymbol{B}\boldsymbol{B}^H\boldsymbol{P} + \boldsymbol{Q}_\varepsilon = \boldsymbol{0} \tag{4.21}$$

This positive definite solution $\boldsymbol{P}_\varepsilon$ has the following properties:

(i) For any $\varepsilon \in (0,1]$, the unique solution $\boldsymbol{P}_\varepsilon > 0$ is such that $\boldsymbol{A}$-$\boldsymbol{BB}^H\boldsymbol{P}_\varepsilon$ is Hurwitz-stable.

(ii) $\lim_{\varepsilon \to 0} \boldsymbol{P}_\varepsilon = 0$.

(iii) $\boldsymbol{P}_\varepsilon$ is continuously differentiable with respect to ε and

$$\frac{d\boldsymbol{P}_\varepsilon}{d\varepsilon} > 0 \quad for \ \ any \quad \varepsilon \in (0, 1] \tag{4.22}$$

Based on the *Lemma 4.2*, low-gain state feedback control laws parameterized in ε is defined by

$$\boldsymbol{u}_L = \boldsymbol{F}_L(\varepsilon)\boldsymbol{x} \tag{4.23}$$

where

$$\boldsymbol{F}_L(\varepsilon) := -\boldsymbol{B}^H\boldsymbol{P}_\varepsilon, \quad \varepsilon \in (0, 1] \tag{4.24}$$

Theorem 4.2 *Consider the system given in Eq.* (4.18)*. Assume that the pair (***A***,* **B***) is stabilizable and the eigenvalues of* **A** *are all located in the closed left-half plane. For any given (arbitrarily large) bounded set* $\chi \subset \mathbb{C}^n$*, there exist an* $\varepsilon^* \in (0,1]$ *such that for all* $\varepsilon \in (0, \varepsilon^*]$*, the equilibrium point* $\boldsymbol{x}= 0$ *of the closed-loop system is locally exponentially stable with* χ *contains in its domain of attraction. Then the family of linear static state feedback laws given in Eq.* (4.23) *solves the problem of semi-global stabilization.*

Proof Let us first consider the state feedback case. Under the state feedback law (4.23), the closed-loop system takes the following form

$$\dot{\boldsymbol{x}} = \boldsymbol{Ax} + \boldsymbol{B}sat\left(-\boldsymbol{B}^H\boldsymbol{P}_\varepsilon\boldsymbol{x}\right) \tag{4.25}$$

When saturations do not occur, define a set as follows

$$\mathscr{L}_\varepsilon \triangleq \left\{\boldsymbol{x} \in C^{2n}:\left|\boldsymbol{B}_j^H\boldsymbol{P}_\varepsilon\boldsymbol{x}\right| \le u_j^{\max} \quad for \quad j = 1, 2, \cdots, m\right\} \tag{4.26}$$

and the closed-loop system is

$$\dot{\boldsymbol{x}} = \left(\boldsymbol{A} - \boldsymbol{BB}^H\boldsymbol{P}_\varepsilon\right)\boldsymbol{x} \tag{4.27}$$

For any given $\varepsilon > 0$, the domain of attraction can be defined as an ellipsoid

$$\mathscr{E}(\boldsymbol{P}_\varepsilon) \triangleq \left\{\boldsymbol{x} \in C^{2n}:\boldsymbol{x}^H\boldsymbol{P}_\varepsilon\boldsymbol{x} \le c\right\} \subseteq \chi \tag{4.28}$$

Let $c > 0$ be

$$c = \min_{j=1,2,\cdots,m} \frac{u_j^{\max}}{\left\{ \boldsymbol{B}_j^H \boldsymbol{P}_\varepsilon \boldsymbol{B}_j \right\}} \tag{4.29}$$

implies that

$$\mathscr{E}(\boldsymbol{P}_\varepsilon) \subseteq \mathscr{L}_\varepsilon \tag{4.30}$$

The given set χ is bounded and $\lim_{\varepsilon \to 0} \boldsymbol{P}_\varepsilon = 0$, so a $\varepsilon*$ exists and satisfies the constraints

$$\chi \subseteq \boldsymbol{E}(\boldsymbol{P}_{\varepsilon^*}, c) \subseteq \boldsymbol{L}_{\varepsilon^*} \tag{4.31}$$

and hence, for all $\varepsilon \in (0, \varepsilon*]$, $\text{sat}(-\boldsymbol{B}^H \boldsymbol{P}_\varepsilon \boldsymbol{x}) = -\boldsymbol{B}^H \boldsymbol{P}_\varepsilon \boldsymbol{x}$. Then the closed-loop system (4.18) behaves linearly and can be written as

$$\dot{\boldsymbol{x}} = \left(\boldsymbol{A} - \boldsymbol{B}\boldsymbol{B}^H \boldsymbol{P}_\varepsilon \right) \boldsymbol{x} \tag{4.32}$$

When the state of the system converges into a smaller set, ε can be made larger so that the feedback gain with a higher level is induced.

Consider a set

$$\boldsymbol{\varepsilon} = \{\varepsilon_0, \varepsilon_1, \ldots, \varepsilon_{\mathcal{N}}\} \quad \varepsilon_i \in \mathbb{R}^+ \ \ and \ \ \varepsilon_i < \varepsilon_{i+1} \quad (i = 0, 1, \ldots, \mathcal{N}) \tag{4.33}$$

where is a real nonnegative integer.

The corresponding ellipsoids yield

$$\boldsymbol{E}(\boldsymbol{P}_0) \supseteq \boldsymbol{E}(\boldsymbol{P}_1) \supseteq \cdots \supseteq \boldsymbol{E}(\boldsymbol{P}_{\mathcal{N}}) \tag{4.34}$$

Based on the sequence of nested ellipsoids, the nested controller is defined as

$$\boldsymbol{u}(\boldsymbol{x}) = \begin{cases} \boldsymbol{u}_0 = -(1+\alpha_0)\boldsymbol{B}^H \boldsymbol{P}_0 \boldsymbol{x} & if \ \ \boldsymbol{x} \in \boldsymbol{E}(\boldsymbol{P}_0) \backslash \boldsymbol{E}(\boldsymbol{P}_1) \\ \boldsymbol{u}_1 = -(1+\alpha_1)\boldsymbol{B}^H \boldsymbol{P}_1 \boldsymbol{x} & if \ \ \boldsymbol{x} \in \boldsymbol{E}(\boldsymbol{P}_1) \backslash \boldsymbol{E}(\boldsymbol{P}_2) \\ \cdots & \cdots \\ \boldsymbol{u}_{\mathcal{N}} = -(1+\alpha_{\mathcal{N}})\boldsymbol{B}^H \boldsymbol{P}_{\mathcal{N}} \boldsymbol{x} & if \ \ \boldsymbol{x} \in \boldsymbol{E}(\boldsymbol{P}_{\mathcal{N}}) \end{cases} \tag{4.35}$$

with the scheduled coefficient of the feedback gain

$$\alpha_i(t) = \int_{t_i}^{t} e^{\frac{\eta_i(t)}{2}\tau} d\tau - 1 \qquad (t_i \leq t < t_{i+1}) \tag{4.36}$$

and

$$\eta_i(t) = \frac{\lambda_{\min}\left[\boldsymbol{Q}_{\varepsilon,i} + \alpha_i \boldsymbol{P}_i \boldsymbol{B}\boldsymbol{B}^H \boldsymbol{P}_i + (1+\alpha_i)\boldsymbol{P}_i^H \boldsymbol{B}\boldsymbol{B}^H \boldsymbol{P}_i\right]}{\|\boldsymbol{P}_i\|} \tag{4.37}$$

t_i represents the time when the state of the system reaches at the boundary of the ith ellipsoid $\boldsymbol{E}(\boldsymbol{P}_i)$. $\lambda_{min}(\bullet)$ denotes the minimum real part of eigenvalues of the a positive definite matrix.

Theorem 4.3 *The nested and gain scheduled controller given in Eq. (4.35) can make the system given in Eq. (4.18) asymptotic stable and the input of the actuators will respect to the constraints.*

Proof At the time $t = t_i$, the state coming from the boundary of $\boldsymbol{E}(\boldsymbol{P}_{i-1})$ reaches at the boundary of $\boldsymbol{E}(\boldsymbol{P}_i)$.For any $\boldsymbol{x} \in \boldsymbol{E}(\boldsymbol{P}_i)$, the closed-loop system under the state feedback law given in Eq. (4.35) may be expressed as

$$\dot{\boldsymbol{x}}(t) = \left[\boldsymbol{A} - (1+\alpha_i)\boldsymbol{B}\boldsymbol{B}^H \boldsymbol{P}_i\right]\boldsymbol{x} \tag{4.38}$$

Consider the Lyapunov function

$$V(\boldsymbol{P}_i, t) = \boldsymbol{x}(t)^H \boldsymbol{P}_i \boldsymbol{x}(t) \tag{4.39}$$

The time-derivative of $V(\boldsymbol{P}_i,t)$ can be evaluated as

$$\begin{aligned}
\dot{V}(\boldsymbol{P}_i, t) &= \boldsymbol{x}^H \boldsymbol{P}_i \dot{\boldsymbol{x}} + \dot{\boldsymbol{x}}^H \boldsymbol{P}_i \boldsymbol{x} \\
&= \boldsymbol{x}^H \boldsymbol{P}_i \left[\boldsymbol{A} - (1+\alpha_i)\boldsymbol{B}\boldsymbol{B}^H \boldsymbol{P}_i\right]\boldsymbol{x} + \boldsymbol{x}^H \left[\boldsymbol{A} - (1+\alpha_i)\boldsymbol{B}\boldsymbol{B}^H \boldsymbol{P}_i\right]^H \boldsymbol{P}_i \boldsymbol{x} \\
&= \boldsymbol{x}^H \left[\boldsymbol{A}^H \boldsymbol{P}_i + \boldsymbol{P}_i \boldsymbol{A} - \boldsymbol{P}_i \boldsymbol{B}\boldsymbol{B}^H \boldsymbol{P}_i - \alpha_i \boldsymbol{P}_i \boldsymbol{B}\boldsymbol{B}^H \boldsymbol{P}_i - (1+\alpha_i)\boldsymbol{P}_i^H \boldsymbol{B}\boldsymbol{B}^H \boldsymbol{P}_i\right]\boldsymbol{x} \\
&= -\boldsymbol{x}^H \left[\boldsymbol{Q}_{\varepsilon,i} + \alpha_i \boldsymbol{P}_i \boldsymbol{B}\boldsymbol{B}^H \boldsymbol{P}_i + (1+\alpha_i)\boldsymbol{P}_i^H \boldsymbol{B}\boldsymbol{B}^H \boldsymbol{P}_i\right]\boldsymbol{x} \\
&\le -\eta_i \boldsymbol{x}^H \boldsymbol{P}_i \boldsymbol{x} \\
&= -\eta_i V(\boldsymbol{P}_i, t)
\end{aligned} \tag{4.40}$$

Hence,

$$\dot{V}(\boldsymbol{P}_i, t) \le e^{-\eta_i t} V(\boldsymbol{P}_i, t) \tag{4.41}$$

For any given time $t \in [t_i, t_{i+1})$, we can obtain the following equation

$$\begin{aligned}
V(\boldsymbol{P}_i, t) &= \boldsymbol{x}(t)^H \boldsymbol{P}_i \boldsymbol{x}(t) \\
&\le \int_{t_i}^{t} e^{-\eta_i(t)\tau} d\tau V(\boldsymbol{P}_i, t_i)
\end{aligned} \tag{4.42}$$

When $t = t_i$, the coefficient α_i is zero and the state satisfies

$$\boldsymbol{x}(t_i) \in \boldsymbol{E}(\boldsymbol{P}_i) \qquad \boldsymbol{x}(t_i) \in \boldsymbol{L}_i \tag{4.43}$$

namely,

$$V(\boldsymbol{P}_i, t_i) = \boldsymbol{x}(t_i)^H \boldsymbol{P}_i \boldsymbol{x}(t_i) \leq c_i \quad and \quad \left|\boldsymbol{B}_j^H \boldsymbol{P}_i \boldsymbol{x}\right| \leq u_j^{\max} (j = 1, 2, \cdots, m) \tag{4.44}$$

Define an intermediate variable as follows

$$\tilde{\boldsymbol{x}}(t) = (1 + \alpha_i)\boldsymbol{x}(t) \tag{4.45}$$

And the Lyapunov function of the intermediate variable can be expressed as

$$\begin{aligned} \tilde{V}(\boldsymbol{P}_i, t) &= \tilde{\boldsymbol{x}}(t)^H \boldsymbol{P}_i \tilde{\boldsymbol{x}}(t) \\ &= (1 + \alpha_i)^2 \boldsymbol{x}(t)^H \boldsymbol{P}_i \boldsymbol{x}(t) \\ &= \int_{t_i}^{t} e^{\eta_i(t)\tau} d\tau V(\boldsymbol{P}_i, t) \end{aligned} \tag{4.46}$$

According to the Eq. (4.42), we can get

$$\begin{aligned} \tilde{V}(\boldsymbol{P}_i, t) &\leq \int_{t_i}^{t} e^{\eta_i(t)\tau} d\tau \int_{t_i}^{t} e^{-\eta_i(t)\tau} d\tau V(\boldsymbol{P}_i, t_i) \\ &= V(\boldsymbol{P}_i, t_i) \end{aligned} \tag{4.47}$$

Hence, the intermediate variable is included in the domain of attraction and will not exceed the saturation region, namely

$$\tilde{\boldsymbol{x}}(t) \in \boldsymbol{E}(\boldsymbol{P}_i) \Rightarrow \tilde{\boldsymbol{x}}(t) \in \boldsymbol{L}_i \tag{4.48}$$

and

$$\left|\boldsymbol{B}_j^H \boldsymbol{P}_i \tilde{\boldsymbol{x}}(t)\right| \leq u_j^{\max} \quad \Rightarrow \quad \left|(1 + \alpha_i)\boldsymbol{B}^H \boldsymbol{P}_i \boldsymbol{x}(t)\right| \leq u_j^{\max} \tag{4.49}$$

So the input of actuators given in Eq. (4.35) will respect to the constraints.

The proof is finished. We can conclude that the coefficient α depends on the convergence rate of the Lyapunov function.

4.2.3 Validation

In this section, a spinning smart beam is utilized to demonstrate effectiveness of the controller designed in the previous section. First, the proposed controller is applied in the case that the only single input is considered. Then a case of multiple-input is simulated with input saturation. The performance of the conventional low gain controller is further studied for comparison.

4.2.3.1 Simulation Model

In this section, a numerical method is used to evaluate the control effect of the gain scheduling controller, and the model uses the Example 3 in Sect. 3.2.1.1.

One end of the beam is attached to a moving base which rotates at a constant angular velocity denoted as Ω and the other end is free, as shown in Fig. 4.1. Rectangular piezoelectric patches of dimensions $54 \times 20 \times 1$ mm are used for actuators, and the material and properties of piezoelectric patches are listed in Table 4.1. The beam is discretized as fifty elements. The stiffness and the mass of the piezoelectric layers are included in the model and the structural damping is neglected in this case.

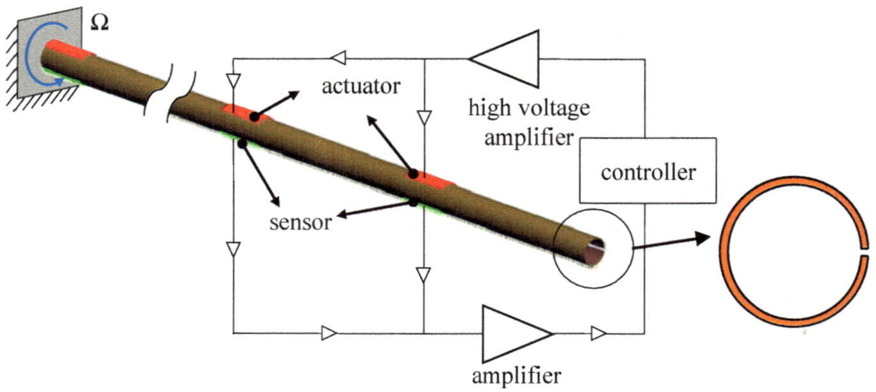

Fig. 4.1 Schematic drawing of a flexible beam with sensors/actuators

Table 4.1 Physical properties of piezoelectric materials

Material properties	PZT
d_{33} ($\times 10^{-12}$ m/V)	300
d_{31} ($\times 10^{-12}$ m/V)	−150
e_{31} (C/m^2)	−7.5
Maximum electric field strength (V/mm)	2000
Young's modulus (GPa)	50
Density (kg/m^3)	7600

4.2.3.2 Analysis with Multiple Input

In order to evaluate the performance of the designed controller in multiple input case, the beam with the first two modes is considered while higher modes are ignored. Two piezoelectric patches are contained in which one is placed on the *y*-directional surface and the other is bonded to the *z*-directional surface of the beam's base.

The voltage of the piezoelectric patch must in the range of [−1000 V, 1000 V]. The following three cases are compared:

Case I: $\mathcal{N} = 0$, $\alpha = 0$ (i.e. the low gain control)
Case II: $\mathcal{N} = 2$, $\alpha = 0$ (i.e. the nested control)
Case III: $\mathcal{N} = 2$, $\alpha \neq 0$ (i.e. the nested control && gain scheduled control)

The initial value of the first two modal coordinates is set to $[1 + i,\ 1-i]^{\mathrm{T}}$, respectively. The initial velocities are set to zero.

Figure 4.2 depicts the history of the parameter α and Fig. 4.3 gives the time history of the control effort of the patches. Figure 4.4 shows control performance comparisons for the displacement of the free end. Because only the first two modes are reserved, the vibration of each direction is dominated by the fundamental bending mode shape. It can be seen that applying the proposed controller achieves the larger level of vibration suppression than the other controllers.

Fig. 4.2 Time-varying parameter α

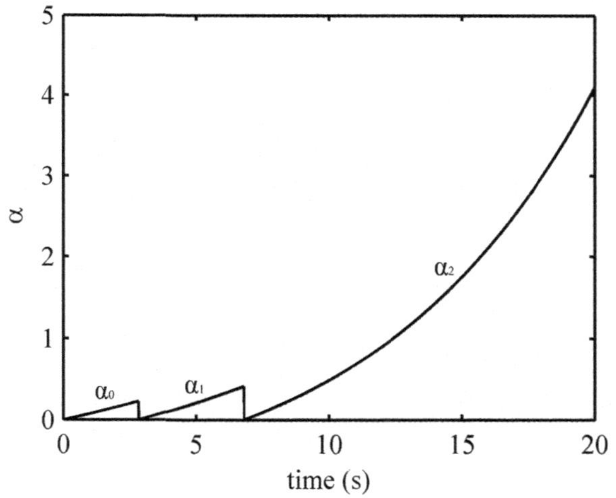

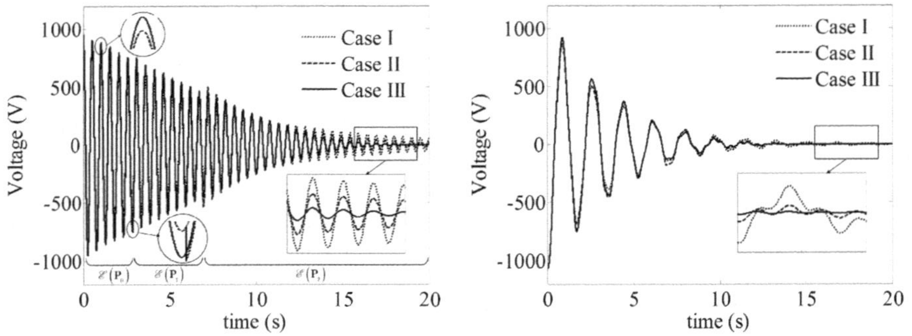

(a) Actuator #1 (b) Actuator #2

Fig. 4.3 Actuator input voltage

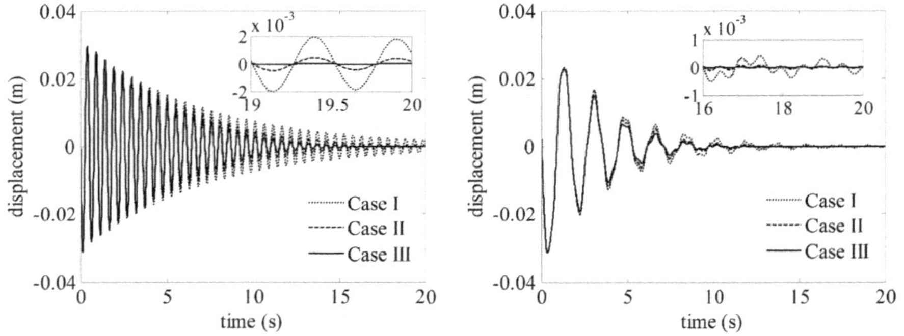

(a) y (b) z

Fig. 4.4 Tip response of beam in body coordinate system

4.3 Gain Scheduled PPF Controller in Complex Mode Space[2]

In order to damp the vibration of the beam containing PZT patches, segmented sensors and actuators with multi–input–multi–output (MIMO) controllers can be used. In this section, a type of classical control law, which is based on positive position feedback is considered. The conventional control law is invariable constant-gain feedback. In the case that the target system is time dependent, a novel variable feedback compensator is designed based on LQR scheme. Brief descriptions of the control law are given as below.

---

[2] Jie Wang, Dongxu Li, Jianping Jiang. Coupled flexural–torsional vibration of spinning smart beams with asymmetric cross sections. Finite Elements in Analysis and Design, 2015,105,16–25.

4.3.1 Gain Scheduled PPF Controller

The positive position feedback controller consists of a second order controller. The second-order system is forced by the position response which is then fed back to give the force input to the structure. The global dynamic equation with positive position feedback control law is given by

$$\begin{aligned} \ddot{\eta} + \overline{C}\dot{\eta} + \Lambda\eta &= B_a G \xi \\ \ddot{\xi} + C_c\dot{\xi} + K_c\xi &= K_c B_s^T \eta \end{aligned} \tag{4.50}$$

The system consists of two equations, one describing the structure and one describing the compensator. G is the feedback gain matrix of PPF, ξ is the compensator state, C_c is damp of the compensator, and K_c is the compensator stiffness matrix. The response, including vibrations and deformations of the beam, is indicated by the modal coordinate η. This response products a voltage on the piezoelectric sensor, which is input to the compensator. The compensator, consisting of a PPF filter, calculates the control voltage. As defined in equation, B_a and B_s are the distribution matrices of actuators and sensors, respectively.

The voltage of actuators is calculated as

$$V_a = -G\xi \tag{4.51}$$

The system in Eq. (4.50) can be expressed as

$$\begin{bmatrix} \ddot{\eta} \\ \ddot{\xi} \end{bmatrix} + \begin{bmatrix} \overline{C} & 0 \\ 0 & C_c \end{bmatrix} \begin{bmatrix} \dot{\eta} \\ \dot{\xi} \end{bmatrix} + \begin{bmatrix} \Lambda & -B_a G \\ -K_c B_s^T & K_c \end{bmatrix} \begin{bmatrix} \eta \\ \xi \end{bmatrix} = 0 \tag{4.52}$$

A state vector x_m containing the modal displacement η, the modal velocity and the compensator is defined as follows

$$x_m = \begin{bmatrix} \eta & \dot{\eta} & \xi & \dot{\xi} \end{bmatrix}^T \tag{4.53}$$

The modal state-space model is built using the new state vector x_m, and can be expressed as a self-governing system

$$\dot{x}_m = A\dot{x}_m \tag{4.54}$$

where $A(t)$ is the closed-loop system matrix, and is given as

$$A = \begin{bmatrix} 0 & I & 0 & 0 \\ -\Lambda & -\overline{C} & B_a G & 0 \\ 0 & 0 & 0 & I \\ K_c B_s^T & 0 & -K_c & -C_c \end{bmatrix} \tag{4.55}$$

The eigenvalues of $\boldsymbol{A}$ give the damped natural frequencies and damping ratios. The system expressed in Eq. (4.54) is stable, if only $\boldsymbol{A}$ is negative definite.

For the target structure studied in this section, the natural frequencies vary with the spinning angular speed. So constant-gain and constant-frequency positive position feedback controller may exceed the stable region of the system. As a consequence, the variable PPF controller with changing frequency and gain is proposed. The stiffness matrix $\boldsymbol{K_c}$, damping matrix $\boldsymbol{C_c}$ and control gain matrix $\boldsymbol{G}$ of the PPF controller are both variant and $\boldsymbol{K_c}$ is consistent with the target structure. The variable PPF compensator extends capability of controlling structural frequencies and damping.

The optimum control gain $\boldsymbol{G}$ is determined based on Linear quadratic regulator (LQR) optimal control theory. In this, the feedback control system is designed to minimize a cost function or a performance index, which is proportional to the required measure of the system's response. The cost function used in this case is given by

$$J = \frac{1}{2}\int_0^{\infty} \left(\boldsymbol{x}^T \boldsymbol{Q}\boldsymbol{x} + \boldsymbol{u}^T \boldsymbol{R}\boldsymbol{u}\right)dt \tag{4.56}$$

where $\boldsymbol{Q}$ is a positive semi definite state penalty matrix and $\boldsymbol{R}$ is a positive definite control penalty matrix. Larger (relatively) elements in $\boldsymbol{Q}$ mean that we demand more vibration suppression ability from the controller. The purpose of the second term in Eq. (4.56) is to account for the effort being expended by the control system, so that small reductions in the output response are not obtained at the expense of physically unreasonable actuator input levels.

The LQR control trajectory minimizes the objective function for arbitrary initial state, namely

$$\min\ J(\boldsymbol{x}_0, \boldsymbol{u}) = J\left(\boldsymbol{x}_0, \boldsymbol{u}_{opt}\right) = \boldsymbol{x}_0^T \boldsymbol{P}\boldsymbol{x}_0 \tag{4.57}$$

where

$$P = P_s + P_u \tag{4.58}$$

$\boldsymbol{P_s}$ and $\boldsymbol{P_u}$ are positive definite matrices and are the solutions to Ricatti equation

$$\begin{aligned} \boldsymbol{A}^T \boldsymbol{P}_s + \boldsymbol{P}_s \boldsymbol{A} + \boldsymbol{Q} = 0 \\ \boldsymbol{A}^T \boldsymbol{P}_u + \boldsymbol{P}_u \boldsymbol{A} + \boldsymbol{G}^T \boldsymbol{R}\boldsymbol{G} = 0 \end{aligned} \tag{4.59}$$

The energy dissipated depends on the initial states which are not always known. In the case that the initial state is assumed to be a random variable, minimization of expected value of J over the set of possible initial states is the same as minimizing *trace*(**P**). Moreover, it is necessary to keep the balance of actuators. So the objective function is defined as

$$\frac{1}{\sqrt[2n]{\prod_{i=1}^{2n} \lambda_i}} trace(\boldsymbol{P}) \quad \rightarrow \quad Min \tag{4.60}$$

where λ_i is the i-th eigenvalue of $\boldsymbol{P}$.

The optimization problem can be described as

$$\left. \begin{array}{l} find\ \boldsymbol{G} \\ \min \dfrac{1}{\sqrt[2n]{\prod_{i=1}^{2n} \lambda_i}} trace(\boldsymbol{P}) \\ s.t\ \ \boldsymbol{A} < 0 \end{array} \right\} \tag{4.61}$$

A genetic algorithm is preferable to some classical and traditional optimization algorithms in obtaining global optimal solutions.

4.3.2 Numerical Applications and Results

In this section, the proposed vibration control theory is applied to the simulation examples. The PPF compensator is used to damp the vibrating beam which is excited by impulse and random loads when the beam is in unchanged spinning angular speed. And the performances of the invariable and variable PPF compensators are compared with a zero-input analysis when the beam is in changing spinning angular speed.

When the angular rate is 0, the stiffness matrix and the damping matrix of the compensator are $diag([12.6, 149.7, 245.0, 400.4, 2116.9])$ and $diag([3.6, 12.2, 15.7, 20.0, 46.0])$, respectively. And the gain matrix is $[160, 95, 10, 1430, 60]$. The natural frequency depends on the rotational angular rate. Therefore, for a rotating beam with a changing speed, the compensator parameters are changing. For example, when the angular rate is 1.2π rad/s, the first five natural frequencies are [0.32, 1.78, 1.78, 3.04, 7.32] Hz. As a result, the stiffness matrix and damping matrix of the compensator are $diag([3.9, 124.4, 124.4, 365.4, 2115.9])$ and $diag([2.0, 11.2, 11.2, 19.1, 46.0])$ respectively, and the gain matrix is $[215, 125, 430, 425, 825]$. The structural damping is set to 0.005.

4.3.2.1 Impact Load

An external impulse load of 0.1 N is assumed to act at the free end of the beam in the z direction for 100 ms duration. The vibration histories of the beam tip without and with the PPF controller (namely open loop and closed loop) are shown in Fig. 4.5

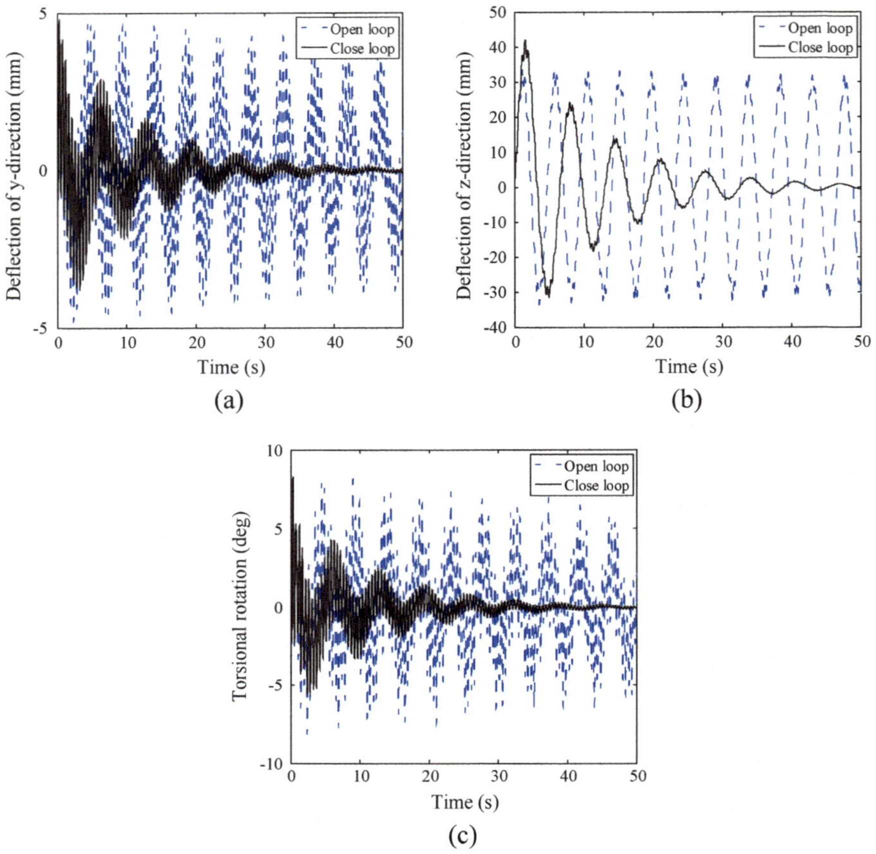

Fig. 4.5 Tip response of the piezoelectric beam with an impact excitation for open-loop and closed-loop system: **(a)** Deflection of *y*-direction, **(b)** Deflection of *z*-direction, **(c)** Torsional rotation

when the spinning speed is 1.2π rad/s. It can be noted that the *z*-directional load excited both deflections of *y*-direction and *z*-direction whose maximum response values are 4.78 and 41.66 mm respectively. After 50 s, the tip deflection of *y*-direction and *z*-direction reduced by 19.9 and 25.0% without the PPF controller while the tip deflection reduced by 98.3 and 98.0% with the PPF controller. The attenuation curve indicates that the vibration levels are reduced faster with the PPF controller.

4.3.2.2 Random Load

The effectiveness of the active control strategy in controlling the response of the beam subject to random load is demonstrated in Fig. 4.6 wherein a random load of *z*-direction is applied to the free end. The random load whose mean value is zero is Gaussian distribution and the amplitude is 0.01 N. The maximum amplitudes of

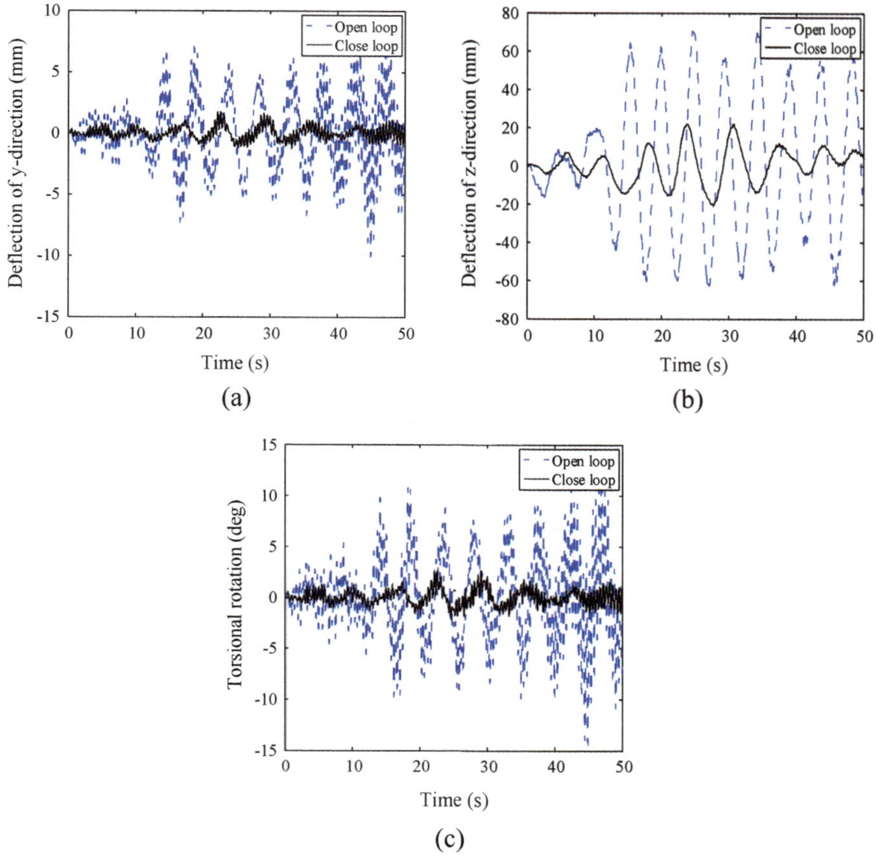

Fig. 4.6 Tip response of the piezoelectric beam subjected to random load for open-loop and closed-loop system: **(a)** Deflection of *y*-direction, **(b)** Deflection of *z*-direction, **(c)** Torsional rotation

three types of deformations are all reduced by almost 80% with PPF compensator compared to the uncontrolled cases.

4.3.2.3 Zero Input Analysis

To demonstrate the advantage of the variable PPF for the active vibration suppression, a case is conducted: the moving base spins in a changed angular speed which changes from 0 to 1.0π rad/s. An initial displacement in *z*-direction is applied to the flexible beam. The invariable PPF controller represents that the compensator is designed when the spinning speed is 0 and used to control the beam regardless the beam's spinning speed. The variable PPF controller is depending on the spinning speed and has variable frequency and control gain. The results of the simulation are shown in Fig. 4.7. It is clear that, from the comparison of curves, the vibration of the flexible

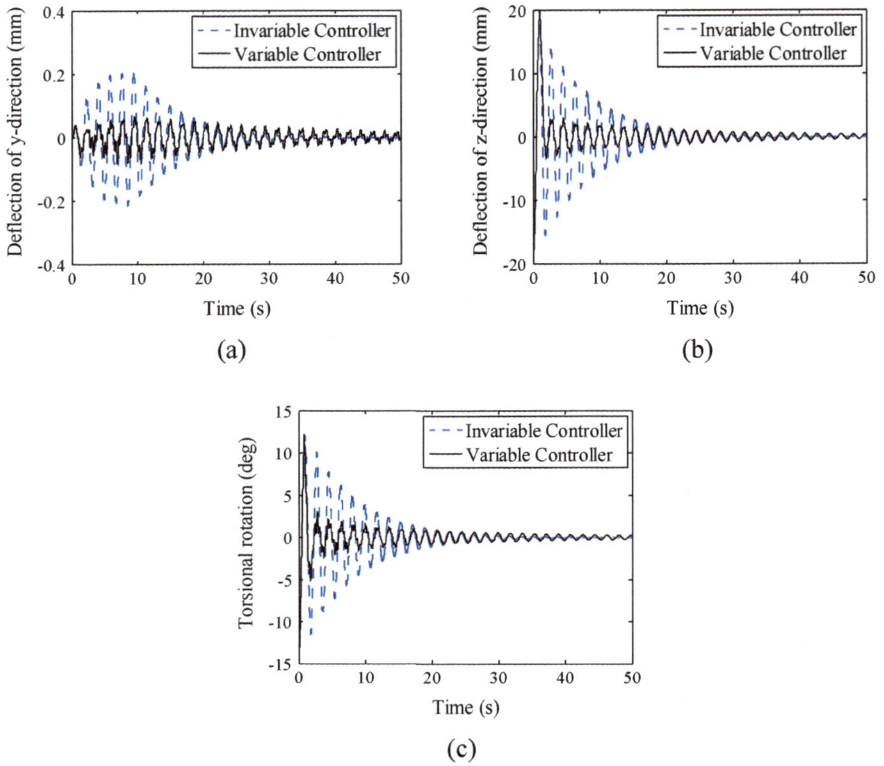

Fig. 4.7 Controlled tip response of the beam with invariable controller and variable controller: **(a)** Deflection of y-direction, **(b)** Deflection of z-direction, **(c)** Torsional rotation

beam with variable PPF controller has significantly been reduced. The system with variable controller has a faster setting time than the invariable controller.

4.4 Sliding Mode Controller in Complex Mode Space

In the previous chapters, the design methods of positive position feedback controller and state feedback controller were proposed for the dynamic system in complex space. However, these two methods need to obtain the accurate dynamic characteristics of the system. The controllers' robustness is poor for uncertainties in the system. According to the control theory in real space, the sliding mode control is a robust control method, which is robust to the non-linear factors and disturbances. Therefore, this section designs a sliding mode controller based on complex modes.

4.4.1 State Transformation

The governing equation of the spinning beam is

$$\begin{aligned} \dot{\boldsymbol{x}} &= \boldsymbol{A}\boldsymbol{x} + \boldsymbol{B}\boldsymbol{u} + \boldsymbol{d} \\ \boldsymbol{V}_s &= \boldsymbol{C}\boldsymbol{x} \end{aligned} \tag{4.62}$$

where

$$\begin{aligned} &\boldsymbol{x} = \begin{bmatrix} \boldsymbol{x}_{1\sim n} \\ \boldsymbol{x}^*_{1\sim n} \end{bmatrix} \quad \boldsymbol{A} = \begin{bmatrix} \boldsymbol{\lambda}_{1\sim n} & \\ & \boldsymbol{\lambda}^*_{1\sim n} \end{bmatrix} \quad \boldsymbol{\Pi}_c = \begin{bmatrix} \boldsymbol{\Pi}_{1\sim n} \\ \boldsymbol{\Pi}^*_{1\sim n} \end{bmatrix} \\ &\boldsymbol{d} = \boldsymbol{\Pi}_c(\boldsymbol{P}_\delta + \boldsymbol{Q}_\delta) \quad \boldsymbol{B} = \begin{bmatrix} \boldsymbol{B}_{1\sim n} \\ \boldsymbol{B}^*_{1\sim n} \end{bmatrix} \quad \boldsymbol{C} = [\boldsymbol{0}\ \gamma^{-1}\boldsymbol{\chi}_s]\boldsymbol{\Phi}_c \end{aligned} \tag{4.63}$$

Assuming that states of the system are all detectable, then the system can be expressed as

$$\dot{\boldsymbol{x}} = \boldsymbol{A}\boldsymbol{x} + \boldsymbol{B}\boldsymbol{u} + \boldsymbol{d} \tag{4.64}$$

Let the state space $\boldsymbol{x}_{1\sim n}$ be composed into two subspaces,

$$\boldsymbol{x}_{1\sim n} = \begin{bmatrix} \boldsymbol{x}_a \\ \boldsymbol{x}_b \end{bmatrix} \quad \boldsymbol{\lambda}_{1\sim n} = \begin{bmatrix} \boldsymbol{A}_a & \boldsymbol{0} \\ \boldsymbol{0} & \boldsymbol{A}_b \end{bmatrix} \quad \boldsymbol{B}_{1\sim n} = \begin{bmatrix} \boldsymbol{B}_a \\ \boldsymbol{B}_b \end{bmatrix} \tag{4.65}$$

Note that $\boldsymbol{u} \in \mathbb{R}^{m\times 1}$, $\boldsymbol{x}_a \in \mathbb{C}^{(n-m)\times 1}$, $\boldsymbol{x}_b \in \mathbb{C}^{m\times 1}$, $\boldsymbol{A}_a \in \mathbb{C}^{(n-m)\times(n-m)}$, $\boldsymbol{A}_b \in \mathbb{C}^{m\times m}$, $\boldsymbol{B}_a \in \mathbb{C}^{(n-m)\times m}$, $\boldsymbol{B}_b \in \mathbb{C}^{m\times m}$. And m is the number of the controls.

Then the Eq. (4.64) can be written as

$$\begin{bmatrix} \dot{\boldsymbol{x}}_a \\ \dot{\boldsymbol{x}}_b \\ \dot{\boldsymbol{x}}^*_a \\ \dot{\boldsymbol{x}}^*_b \end{bmatrix} = \begin{bmatrix} \boldsymbol{A}_a & \boldsymbol{0} & \boldsymbol{0} & \boldsymbol{0} \\ \boldsymbol{0} & \boldsymbol{A}_b & \boldsymbol{0} & \boldsymbol{0} \\ \boldsymbol{0} & \boldsymbol{0} & \boldsymbol{A}^*_a & \boldsymbol{0} \\ \boldsymbol{0} & \boldsymbol{0} & \boldsymbol{0} & \boldsymbol{A}^*_b \end{bmatrix} \begin{bmatrix} \boldsymbol{x}_a \\ \boldsymbol{x}_b \\ \boldsymbol{x}^*_a \\ \boldsymbol{x}^*_b \end{bmatrix} + \begin{bmatrix} \boldsymbol{B}_a \\ \boldsymbol{B}_b \\ \boldsymbol{B}^*_a \\ \boldsymbol{B}^*_b \end{bmatrix} \boldsymbol{u} + \boldsymbol{d} \tag{4.66}$$

The state transformation is defined as

$$\boldsymbol{x}_{1\sim n} = \boldsymbol{T}\boldsymbol{y} \quad \boldsymbol{x}^*_{1\sim n} = \boldsymbol{T}^*\boldsymbol{y}^* \tag{4.67}$$

where

$$\boldsymbol{y} = \begin{bmatrix} \boldsymbol{y}_a \\ \boldsymbol{y}_b \end{bmatrix} \quad \boldsymbol{y}^* = \begin{bmatrix} \boldsymbol{y}_a^* \\ \boldsymbol{y}_b^* \end{bmatrix}$$
$$\boldsymbol{T} = \begin{bmatrix} \boldsymbol{I}_{n-m} & \boldsymbol{B}_a \boldsymbol{B}_b^{-1} \\ \boldsymbol{0} & \boldsymbol{I}_m \end{bmatrix} \quad \boldsymbol{T}^* = \begin{bmatrix} \boldsymbol{I}_{n-m} & \boldsymbol{B}_a^* \left(\boldsymbol{B}_b^*\right)^{-1} \\ \boldsymbol{0} & \boldsymbol{I}_m \end{bmatrix} \tag{4.68}$$

Substituting Eq. (4.67) into Eq. (4.66), and the system can be expressed in $\boldsymbol{y}$-coordinate as

$$\dot{\boldsymbol{y}} = \hat{\boldsymbol{A}} \boldsymbol{y} + \hat{\boldsymbol{B}} \boldsymbol{u}$$
$$\dot{\boldsymbol{y}}^* = \hat{\boldsymbol{A}}^* \boldsymbol{y}^* + \hat{\boldsymbol{B}}^* \boldsymbol{u} \tag{4.69}$$

where

$$\hat{\boldsymbol{A}} = \begin{bmatrix} \boldsymbol{A}_{11} & \boldsymbol{A}_{12} \\ \boldsymbol{A}_{21} & \boldsymbol{A}_{22} \end{bmatrix} \quad \hat{\boldsymbol{B}} = \begin{bmatrix} \boldsymbol{0} \\ \boldsymbol{B}_b \end{bmatrix}$$
$$\hat{\boldsymbol{A}}^* = \begin{bmatrix} \boldsymbol{A}_{11}^* & \boldsymbol{A}_{12}^* \\ \boldsymbol{A}_{21}^* & \boldsymbol{A}_{22}^* \end{bmatrix} \quad \hat{\boldsymbol{B}}^* = \begin{bmatrix} \boldsymbol{0} \\ \boldsymbol{B}_b^* \end{bmatrix} \tag{4.70}$$

The equation of the system can be obtained by stacking the basis vectors up to form a matrix

$$\begin{bmatrix} \dot{\boldsymbol{y}}_1 \\ \dot{\boldsymbol{y}}_1^* \\ \dot{\boldsymbol{y}}_2 \\ \dot{\boldsymbol{y}}_2^* \end{bmatrix} = \begin{bmatrix} \boldsymbol{A}_{11} & \boldsymbol{0} & \boldsymbol{A}_{12} & \boldsymbol{0} \\ \boldsymbol{0} & \boldsymbol{A}_{11}^* & \boldsymbol{0} & \boldsymbol{A}_{12}^* \\ \boldsymbol{A}_{21} & \boldsymbol{0} & \boldsymbol{A}_{22} & \boldsymbol{0} \\ \boldsymbol{0} & \boldsymbol{A}_{21}^* & \boldsymbol{0} & \boldsymbol{A}_{22}^* \end{bmatrix} \begin{bmatrix} \boldsymbol{y}_1 \\ \boldsymbol{y}_1^* \\ \boldsymbol{y}_2 \\ \boldsymbol{y}_2^* \end{bmatrix} + \begin{bmatrix} \boldsymbol{0} \\ \boldsymbol{0} \\ \boldsymbol{B}_b \\ \boldsymbol{B}_b^* \end{bmatrix} \boldsymbol{u} + \boldsymbol{d} \tag{4.71}$$

In order to design the switching function, the number of nonzero rows of factors of $\boldsymbol{u}$ should be equaled to m which is the number of controls. Hence, another state transformation must be performed. Specify

$$\begin{bmatrix} \boldsymbol{y}_b \\ \boldsymbol{y}_b^* \end{bmatrix} = \overline{\boldsymbol{T}}_m \begin{bmatrix} \overline{\boldsymbol{y}}_b \\ \overline{\boldsymbol{y}}_b^* \end{bmatrix} \quad \begin{bmatrix} \boldsymbol{y}_a \\ \boldsymbol{y}_a^* \\ \boldsymbol{y}_b \\ \boldsymbol{y}_b^* \end{bmatrix} = \overline{\boldsymbol{T}} \begin{bmatrix} \overline{\boldsymbol{y}}_a \\ \overline{\boldsymbol{y}}_a^* \\ \overline{\boldsymbol{y}}_b \\ \overline{\boldsymbol{y}}_b^* \end{bmatrix} \tag{4.72}$$

where

$$\overline{\boldsymbol{T}}_m = \begin{bmatrix} \boldsymbol{I}_m & \boldsymbol{B}_b \left(\boldsymbol{B}_b^*\right)^{-1} \\ \boldsymbol{0} & \boldsymbol{I}_m \end{bmatrix} \quad \overline{\boldsymbol{T}} = \begin{bmatrix} \boldsymbol{I}_{n-m} & & \\ & \boldsymbol{I}_{n-m} & \\ & & \boldsymbol{T}_m \end{bmatrix} \tag{4.73}$$

Thus the governing equation of the system is

$$\begin{bmatrix} \dot{\overline{y}}_a \\ \dot{\overline{y}}_a^* \\ \dot{\overline{y}}_b \\ \dot{\overline{y}}_b^* \end{bmatrix} = \overline{T}^{-1} \begin{bmatrix} A_{11} & 0 & A_{12} & 0 \\ 0 & A_{11}^* & 0 & A_{12}^* \\ A_{21} & 0 & A_{22} & 0 \\ 0 & A_{21}^* & 0 & A_{22}^* \end{bmatrix} \overline{T} \begin{bmatrix} \overline{y}_a \\ \overline{y}_a^* \\ \overline{y}_b \\ \overline{y}_b^* \end{bmatrix} + \begin{bmatrix} 0 \\ 0 \\ 0 \\ B_b^* \end{bmatrix} u \tag{4.74}$$

The above system is a complex state space with m inputs. Partition the state vector in the obvious manner

$$\begin{bmatrix} \dot{\overline{y}}_a \\ \dot{\overline{y}}_a^* \\ \dot{\overline{y}}_b \\ \hline \dot{\overline{y}}_b^* \end{bmatrix} = \left[\begin{array}{c|c} \bar{A}_{11} & \bar{A}_{12} \\ \hline \bar{A}_{21} & \bar{A}_{22} \end{array} \right] \begin{bmatrix} \overline{y}_1 \\ \overline{y}_1^* \\ \overline{y}_2 \\ \hline \overline{y}_2^* \end{bmatrix} + \begin{bmatrix} 0 \\ 0 \\ 0 \\ \hline \mathbf{B}_b^* \end{bmatrix} \mathbf{u} \tag{4.75}$$

4.4.2 Sliding Surface Vector Design

In general, conventional systems have real state vectors whose coefficients are also real. The sliding manifold is the linear combination of real state vectors. In this research, state vectors and their coefficients are all complex. Though, the switch function maintains real.

Let the switching function for the system described in Eq. (4.71) be

$$\sigma := \begin{bmatrix} S & S^* \end{bmatrix} \begin{bmatrix} y \\ y^* \end{bmatrix} \tag{4.76}$$

where S is a complex constant matrix and $S \in \mathbb{C}^{m \times n}$. The definition of conjugate matrices ensures the switching function is in real number space, namely $\sigma \in \mathbb{R}^{m}$. Partition the matrix S as $S = [S_a, S_b]$, and $S_a \in \mathbb{C}^{m\times(n-m)}$, $S_b \in \mathbb{C}^{m\times m}$. Then the switching function can be written as

$$\sigma = \begin{bmatrix} S_a & S_a^* & S_b & S_b^* \end{bmatrix} \begin{bmatrix} y_a \\ y_a^* \\ y_b \\ y_b^* \end{bmatrix} = 2\left(S_a y_a + S_b y_b\right)^R \tag{4.77}$$

After transformation, the switching function can be expressed as

$$\sigma = \begin{bmatrix} S_a & S_a^* & S_b & S_b^* \end{bmatrix} \overline{T} \begin{bmatrix} \overline{y}_a \\ \overline{y}_a^* \\ \overline{y}_b \\ \overline{y}_b^* \end{bmatrix}$$
$$= \begin{bmatrix} S_a & S_a^* & S_b & S_b B_b \left(B_b^*\right)^{-1} + S_b^* \end{bmatrix} \begin{bmatrix} \overline{y}_a \\ \overline{y}_a^* \\ \overline{y}_b \\ \overline{y}_b^* \end{bmatrix} \tag{4.78}$$

By assigning $\sigma(t) = 0$, the equivalent dynamic system in the sliding mode can be obtained

$$\begin{bmatrix} \dot{\overline{y}}_a \\ \dot{\overline{y}}_a^* \\ \dot{\overline{y}}_b \end{bmatrix} = \left(\overline{A}_{11} - \overline{A}_{12} K\right) \begin{bmatrix} \overline{y}_a \\ \overline{y}_a^* \\ \overline{y}_b \end{bmatrix} \tag{4.79}$$

where

$$K = \left[S_b B_b \left(B_b^*\right)^{-1} + S_b^*\right]^{-1} \begin{bmatrix} S_a & S_a^* & S_b \end{bmatrix} \tag{4.80}$$

For controllable system in Eq. (4.79), $2n$-m eigenvalues of the sliding motion may be located at our will by a proper choice of K. Choose $S_a \in \mathbb{C}^{m\times(n-m)}$ and $S_b \in \mathbb{C}^{m\times m}$ so that $(\overline{A}_{11} - \overline{A}_{12} K)$ has prescribed eigenvalues in the open left half plane.

Let us denote by u_{eq} (the equivalent control) the control which obtains while the trajectory remains in the manifold $\dot{\sigma}(t) = 0$. Then, u_{eq} is defined by

$$\mathbf{u}_{eq} = -\left(\begin{bmatrix} \mathbf{S} & \mathbf{S}^* \end{bmatrix} \begin{bmatrix} \hat{B} \\ \hat{B}^* \end{bmatrix}\right)^{-1} \begin{bmatrix} \mathbf{S} & \mathbf{S}^* \end{bmatrix} \begin{bmatrix} \hat{A} & \mathbf{0} \\ \mathbf{0} & \hat{A}^* \end{bmatrix} \begin{bmatrix} \mathbf{y} \\ \mathbf{y}^* \end{bmatrix}$$
$$= -\left(\mathbf{S}_b \mathbf{B}_b + \mathbf{S}_b^* \mathbf{B}_b^*\right)^{-1}$$
$$\begin{bmatrix} \mathbf{S}_a \mathbf{A}_{11} + \mathbf{S}_b \mathbf{A}_{21} & \mathbf{S}_a^* \mathbf{A}_{11}^* + \mathbf{S}_b^* \mathbf{A}_{21}^* & \mathbf{S}_a \mathbf{A}_{12} + \mathbf{S}_b \mathbf{A}_{22} & \mathbf{S}_a^* \mathbf{A}_{12}^* + \mathbf{S}_b^* \mathbf{A}_{22}^* \end{bmatrix} \begin{bmatrix} \mathbf{y} \\ \mathbf{y}^* \end{bmatrix}$$
$$= -\left[(\mathbf{S}_b \mathbf{B}_b)^R\right]^{-1} \left(S \hat{A} y\right)^R \tag{4.81}$$

It can be seen that a nonzero determinant of the product matrix $(S_b B_b)$ is needed.

4.4.3 Controller Design Under Input Saturation

In most of the practical systems, the actuator capacity is limited. For the system studied in the present paper, the input voltages of the piezoelectric patches, denoted by $\boldsymbol{u}$, should respect its boundaries. In this section, the control design under actuator constraints is considered

$$\begin{aligned}\dot{\boldsymbol{y}} &= \hat{\boldsymbol{A}}\boldsymbol{y} + \hat{\boldsymbol{B}} sat(\boldsymbol{u}) \\ \dot{\boldsymbol{y}}^* &= \hat{\boldsymbol{A}}^*\boldsymbol{y}^* + \hat{\boldsymbol{B}}^* sat(\boldsymbol{u})\end{aligned} \tag{4.82}$$

where matrices $\hat{\boldsymbol{A}}$, $\hat{\boldsymbol{B}}$, $\hat{\boldsymbol{A}}^*$, and $\hat{\boldsymbol{B}}^*$ are given in Eq. (4.70).

Lemma 4.3 *For a physically reliable system presented in Eq. (4.82), the state vector* $\boldsymbol{y}$*is limited and there exists a known positive value, such that*

$$\left\| \left[\left(S\hat{\boldsymbol{B}} \right)^R \right]^{-1} \left(S\hat{\boldsymbol{A}}y \right)^R \right\| \leq \varepsilon \tag{4.83}$$

Lemma 4.4 *For a limited input denoted by* $\boldsymbol{u}$*, the nonlinearity sat(*$\boldsymbol{u}$*) satisfies.*

$$\boldsymbol{u}^T sat(\boldsymbol{u}) \geq \gamma \boldsymbol{u}^T \boldsymbol{u} \tag{4.84}$$

where $\gamma = \min\{\gamma_1, \ldots, \gamma_m\}$, and γ_j is defined as follows

$$\gamma_j u_j^2 \leq u_j sat\left(u_j\right) \qquad (j = 1, \ldots, m) \tag{4.85}$$

Lemma 4.5 *For the nonlinear input satisfied Eq. (4.84), there exists a continuous function* $\varphi(\cdot)$*:* $\mathbb{R}_+ \to \mathbb{R}_+$*,* $\varphi(0) = 0$*,* $\varphi(p) \geq 0$ *for* $p \geq 0$*, such that, for all* $q \geq 0$*,*

$$\gamma \boldsymbol{u}^T \boldsymbol{u} \geq q\varphi(q) \tag{4.86}$$

Control law to ensure asymptotical stability of the sliding mode takes the form

$$\boldsymbol{u}(t) = -\frac{\left(\hat{\boldsymbol{B}}^T \boldsymbol{S}^T\right)^R \boldsymbol{\sigma}(t)}{\left\| \left(\hat{\boldsymbol{B}}^T \boldsymbol{S}^T\right)^R \boldsymbol{\sigma}(t) \right\|} \varphi(q) \tag{4.87}$$

where $\varphi(q) = \delta q$ with $\delta \geq 1/\gamma$, and function $q(\bullet)$ is defined as

$$q(\boldsymbol{y}, t) = \eta \left\| \left[\left(S\hat{\boldsymbol{B}} \right)^R \right]^{-1} \left(S\hat{\boldsymbol{A}}y \right)^R \right\| \qquad \eta > 1 \tag{4.88}$$

Theorem 4.4 *For the system in Eq. (4.82)with saturation input constraints and the sliding surface in Eq. (4.76), the controller in Eq. (4.87)can drive the closed-loop system onto the sliding surface* $\boldsymbol{\sigma} = 0$.

Proof A Lyapunov functional candidate is constructed as

$$V = \frac{1}{2}\sigma^T \sigma \tag{4.89}$$

Taking the time derivative of V and substituting Eq. (4.76) into the time derivative,

$$\begin{aligned}
\dot{V} &= \boldsymbol{\sigma}^T \dot{\boldsymbol{\sigma}} \\
&= \boldsymbol{\sigma}^T \left(\boldsymbol{S}\hat{\boldsymbol{A}}\dot{\boldsymbol{y}} + \boldsymbol{S}^* \hat{\boldsymbol{A}}^* \dot{\boldsymbol{y}}^* \right) \\
&= \boldsymbol{\sigma}^T \left[\boldsymbol{S}\hat{\boldsymbol{A}}\boldsymbol{y} + \boldsymbol{S}\hat{\boldsymbol{B}} \cdot sat(\boldsymbol{u}) + \boldsymbol{S}^* \hat{\boldsymbol{A}}^* \boldsymbol{y}^* + \boldsymbol{S}^* \hat{\boldsymbol{B}}^* \cdot sat(\boldsymbol{u}) \right] \\
&= 2\boldsymbol{\sigma}^T \left(\boldsymbol{S}\hat{\boldsymbol{A}}\boldsymbol{y} \right)^R + 2\boldsymbol{\sigma}^T \left(\boldsymbol{S}\hat{\boldsymbol{B}} \right)^R \cdot sat(\boldsymbol{u}) \\
&= 2\boldsymbol{\sigma}^T \left(\boldsymbol{S}\hat{\boldsymbol{B}} \right)^R \left[\left(\boldsymbol{S}\hat{\boldsymbol{B}} \right)^R \right]^{-1} \left(\boldsymbol{S}\hat{\boldsymbol{A}}\boldsymbol{y} \right)^R + 2\boldsymbol{\sigma}^T \left(\boldsymbol{S}\hat{\boldsymbol{B}} \right)^R \cdot sat(\boldsymbol{u})
\end{aligned} \tag{4.90}$$

From Eqs. (4.84) and (4.86),

$$\sigma^T \left(\boldsymbol{S}\hat{\boldsymbol{B}} \right)^R \cdot sat(\boldsymbol{u}) \leq - \left\| \sigma^T \left(\boldsymbol{S}\hat{\boldsymbol{B}} \right)^R \right\| q(\boldsymbol{y}, t) \tag{4.91}$$

Substituting Eq. (4.91) into Eq. (4.90), then

$$\begin{aligned}
\dot{V} &= 2\sigma^T (\boldsymbol{S}\boldsymbol{B})^R \left[\left(\boldsymbol{S}\hat{\boldsymbol{B}} \right)^R \right]^{-1} \left(\boldsymbol{S}\hat{\boldsymbol{A}}y \right)^R + 2\sigma^T \left(\boldsymbol{S}\hat{\boldsymbol{B}} \right)^R \cdot sat(\boldsymbol{u}) \\
&\leq 2\sigma^T \left(\boldsymbol{S}\hat{\boldsymbol{B}} \right)^R \left[\left(\boldsymbol{S}\hat{\boldsymbol{B}} \right)^R \right]^{-1} \left(\boldsymbol{S}\hat{\boldsymbol{A}}y \right)^R - 2 \left\| \sigma^T \left(\boldsymbol{S}\hat{\boldsymbol{B}} \right)^R \right\| q(\boldsymbol{y}, t) \\
&\leq 2 \left\| \sigma^T \left(\boldsymbol{S}\hat{\boldsymbol{B}} \right)^R \right\| \left[\left(\boldsymbol{S}\hat{\boldsymbol{B}} \right)^R \right]^{-1} \left(\boldsymbol{S}\hat{\boldsymbol{A}}y \right)^R - 2\eta \left\| \sigma^T \left(\boldsymbol{S}\hat{\boldsymbol{B}} \right)^R \right\| \left\| \left[\left(\boldsymbol{S}\hat{\boldsymbol{B}} \right)^R \right]^{-1} \left(\boldsymbol{S}\hat{\boldsymbol{A}}y \right)^R \right\| \\
&\leq 2(1-\eta) \left\| \sigma^T \left(\boldsymbol{S}\hat{\boldsymbol{B}} \right)^R \right\| \left[\left(\boldsymbol{S}\hat{\boldsymbol{B}} \right)^R \right]^{-1} \left(\boldsymbol{S}\hat{\boldsymbol{A}}y \right)^R \\
&< 0
\end{aligned} \tag{4.92}$$

Clearly, since $V(\boldsymbol{\sigma})$ is positive definite and $\dot{V}(\boldsymbol{\sigma})$ is negative definite, the system will be driven to the sliding manifold $\boldsymbol{\sigma} = 0$ according to Lyapunov's second method.

4.4.4 Numerical Applications and Results

The simulation example uses the model in Sect. 4.2.3.1. In order to reduce the degrees of freedom of the system, the first two elastic modes are reserved and higher order modes are ignored. Piezoelectric patch, which act as actuators, are placed on the surface of the beam close to the fixed end. The allowable voltage of the piezoelectric actuators used in the system is from -1000 to $+ 1000$ V. The performance of the control system in attenuating the unwanted vibrations was investigated by setting the initial condition of the beam and observing the tip response with or without the controller. With initial condition $y(0) = [1 + i, 0]^{\mathrm{T}}$ the system is simulated with maximum sampling time 0.001 s. The results of the simulation are shown in Figs. 4.8, 4.9, 4.10 and 4.11.

Both viscous damping and structural damping are neglected in the analysis. Hence, the beam experiences undamped oscillation without control, as shown in Figs. 4.8

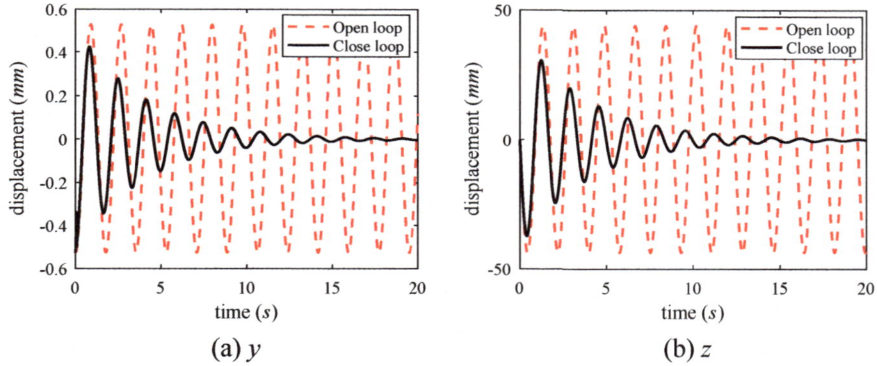

Fig. 4.8 Tip response of the beam in the body coordinate **(a)** y and **(b)** z

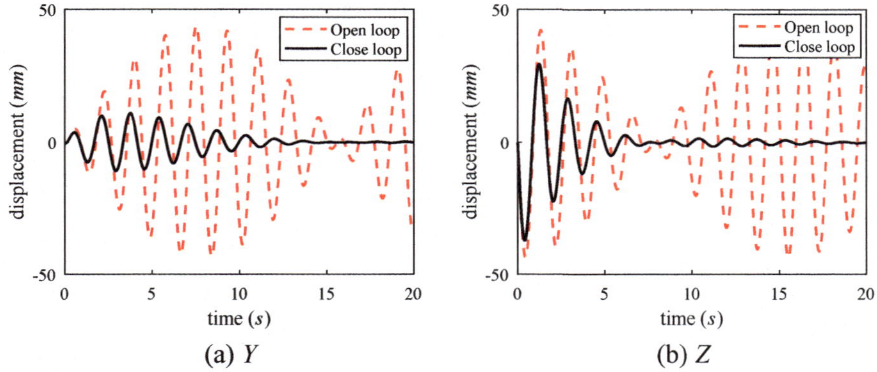

Fig. 4.9 Tip response of the beam in the inertial coordinate **(a)** Y and **(b)** Z

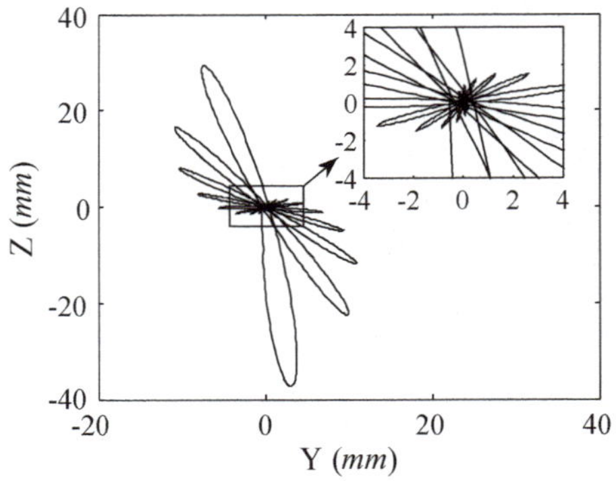

Fig. 4.10 Transverse track of the tip in the inertial coordinate with SMC

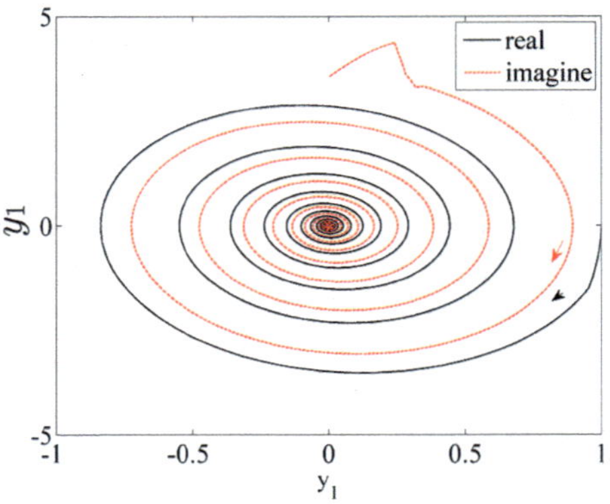

Fig. 4.11 Phase plane of the first complex mode with SMC

and 4.9. Figure 4.8 plots the time response of the free end of the beam with SMC in the body coordinate. Figure 4.9 shows the tip response of the beam with SMC under the inertial coordinate. Figure 4.10 curves the transverse track of the tip. The transverse displacement converges to zero with respect to time. It is clear that the vibration of the flexible beam has significantly been reduced.

The state of the first complex mode with SMC is shown in Fig. 4..4.11, which shows that the SMC achieves a good performance in the vibration attenuation.

4.5 Summary

This chapter proposes the design methods of the state feedback controller, the positive position feedback controller and the sliding mode controller for systems in the complex modal space. Numerical simulations have been conducted to verify the effectiveness of the methods. The main conclusions are as follows:

(1) The time-varying low–high gain controller, which was designed based on nested control and gain scheduling control, shows good performance under

both single-input and multiple-input conditions, and improves the attenuation rate of the elastic vibration of the flexible beam.

(2) Effectiveness of the variable gain positive position feedback controller has been validated by comparison with the open-looped system under various forces including pulse, random and initial displacement cases.

(3) The sliding mode controller in complex modal space has been proposed for systems with uncertainties. And the nonlinear controller has also considered the saturation of actuators.

References

1. Neto, M. A., Ambrosio, J. A., Roseiro, L. M., et al. (2013). Active vibration control of spatial flexible multibody systems. *Multibody System Dynamics, 30*(1), 13–35.
2. Huang, Y-A., & Deng, Z-C. (2005) Decentralized sliding mode control for a spacecraft flexible appendage based on finite element method. *Chinese Journal of Aeronautics, 18*(3), 230–6.
3. Okubo, H., Yano, H., & Itoh, T. (2013) Semi-active vibration suppression for smart structures with sliding-mode control. *Journal of Intelligent Material Systems and Structures*, 1045389X13512907.
4. Bessa, W., de Paula, A., & Savi, M. (2013). Adaptive fuzzy sliding mode control of smart structures. *The European Physical Journal Special Topics, 222*(7), 1541–1551.
5. Sangpet, T., Kuntanapreeda, S., & Schmidt, R. (2014) Adaptive vibration control of piezoactuated euler-bernoulli beams using infinite-dimensional lyapunov method and high-order sliding-mode differentiation. *Journal of Engineering*, 2014.
6. Zhang, T., Li, H. G., & Cai, G. P. (2013). Hysteresis identification and adaptive vibration control for a smart cantilever beam by a piezoelectric actuator. *Sensors and Actuators A: Physical, 203*, 168–75.
7. Balas, M. J. (1978). Active control of flexible systems. *Journal of Optimization theory and Applications, 25*(3), 415–436.
8. Meirovitch, L., & Shenhar, J. (1984). NASA Contractor Report 3760.
9. Meyer, J. L., Harrington, W. B., Agrawal, B. N., et al. (1998). Vibration suppression of a spacecraft flexible appendage using smart material. *Smart Materials and Structures, 7*(1), 95.
10. Jiang, J-P., & Li, D-X. (2010). Robust H_∞ vibration control for smart solar array structures. Journal of Vibration and Control, *17*, 505–515.
11. Koofigar, H. R., & Amelian, S. (2012). Active vibration suppression in smart structures subjected to model uncertainties and environmental disturbances: an adaptive approach. *Journal of Vibration and Control, 19*(13), 2046–2053.
12. Changjian, S., Xianmin, Z., & Yunwen, S. (2000). Complex mode active vibration control of high-speed flexible linkage mechanisms. *Journal of Sound and Vibration, 234*(3), 491–506.
13. Zhang, X., Shao, C., Li, S., et al. (2001). Robust H_∞ vibration control for flexible linkage mechanism systems with piezoelectric sensors and actuators. *Journal of Sound and Vibration, 243*(1), 145–155.
14. Gohberg, I., Lancaster, P., & Rodman, L. (1986). On hermitian solutions of the symmetric algebraic riccati equation. *SIAM Journal on Control and Optimization, 24*(6), 1323–1334.

Chapter 5
Optimal Variable Amplitudes Input Shaping Control for Slew Maneuver of Flexible Spacecraft

5.1 Introduction

During the attitude maneuver of the flexible spacecraft, the attitude movement of the spacecraft body will excite the elastic vibration of the flexible attachment [1–3]. In return, the vibration would induce attitude oscillation after maneuver, which will downgrade the performance of payloads in spacecraft. Besides, in order to execute station keeping and attitude maneuvers, spacecraft often contain propellant tanks with large quantities of liquid fuel. And the fuel slosh excited by the main platform would influence the attitude of the spacecraft [4–7]. In addition, the vibration of the flexible appendages and fuel slosh will change the physical properties of the spacecraft [4, 8]. And the fuel consumption over time would introduce a further uncertainty in the satellite model description. And the elastic vibration of the appendage after the maneuver will cause the attitude jitter. Therefore, attitude maneuver path and attitude stability theory are very important to the research of flexible spacecraft attitude maneuver.

For more than three decades, considerable attention has been paid to the investigation of the maneuver control of spacecraft. Proposed methods include the trajectory optimization, in which a path is optimized to cause the minimum vibration of the flexible appendage, and the feedback control technique, in which the induced vibration of the appendage is damped out by positive or active control theories. Among the former techniques, the input shaping method is one of effective theories to reduce elastic vibration of flexible appendages.

Input shaping, which was first proposed by OJM Smith, known as posicast control [9], is a feed-forward technique that seeks to reduce residual vibrations of flexible systems by modification of the input. However, the accuracy of this method depends on the exact parameters of the system, including natural frequencies and damping ratios [10]. In order to improve the robustness, various approaches were proposed

© The Author(s), under exclusive license to Springer Nature Singapore Pte Ltd. 2022
J. Wang and D.-X. Li, *Rigid-Flexible Coupling Dynamics and Control of Flexible Spacecraft with Time-Varying Parameters*,
https://doi.org/10.1007/978-981-16-5097-0_5

such as Zero-Vibration and Derivative (ZVD) shaper, Extra Insensitivity (EI) [11] approach and Specified Insensitivity (SI) concept [12]. And the method was extended to multiple mode input shaper method [13, 14].

Furthermore, the input shaping method was extended to time-varying systems. Magee and Book [15, 16] developed a modified command shaping technique for systems with varying natural frequencies. Pao and Singhose [17] compared variable-amplitude input shaping to other shaped commands. Cho and Park [18] used the state transition matrix to compute the appropriate time-varying input sequence. But the paper neglected the coupling effects between flexible modes. Lee and Park [19] proposed a new shaping scheme based on the modal-filtered impulse response concept. It can be applied to nonlinear systems by linearizing the dynamic equations and then express the governing equations in state-space style. Otsuki et al. [20] developed a robust command shaping method which is independent of the schedule of a system's change. In order to simultaneous control of position target and vibration, they proposed a synthesis procedure for a planetary rover [21]. The amplitude of control input was adjusted with a gain to reach the desired target. The method is capable of controlling systems with single input and a target with one degree. For systems with targets of more degrees, for example, the attitude angles of spacecraft, the method is not suitable.

Because of its well-proven efficiency and effectiveness, the input shaping is widely used in various branches of engineering, such as spacecraft maneuver [22–26], manipulator [27, 28], aircraft control [29], vibration control [30, 31] and transportation mechanical systems. In the field of spacecraft maneuver, the input shaping method is applied to modify trajectory or design an appropriate control profile to reduce the residual vibration [32–39]. Gasbarri et al. [37] proposed a nonlinear attitude controller to reduce the rigid-flexible coupling effect, and minimize the vibration of the flexible structure by designing an input shaping attitude maneuver algorithm. Singhose [32] designed on–off command profiles for a single-axial slew and effectively limited the deflection of flexible appendage. In general, the method is combined with a feedback method to form a closed-loop system. Sung [33] combined the input shaping technique with the sliding-mode control. Parman [34, 35] combined constant-amplitude shaped inputs and proportional derivative (PD) based feedback input to control the rest-to-rest attitude maneuvers of a spacecraft with flexible solar panels. Zhang [36] applied the input shaper method to fast attitude maneuver of flexible spacecraft. Gasbarri et al. [37] modified the reference attitude trajectory with command shaping in a fast re-orienting maneuver. Na et al. [40] studied the vibration of the flexible solar wing during the orbital maneuvering of the spacecraft based on the input shaping technology, carried out ground tests, and demonstrated the engineering applicability by numerical simulation.

Previous researches were mostly focused on the application to systems with uncertainties or varying frequencies. Few research works were intended to address the maneuver of spacecraft with the time-varying parameters. The designed attitude maneuvering strategy has a good robustness when the parameters are accurate or have small changes and uncertainties. These shapers are not suitable when the parameters in the spacecraft dynamics equation vary obviously.

This chapter studies the attitude maneuver of the flexible spacecraft with time-varying parameters. The main purpose is to propose an optimal design strategy for attitude maneuver, which not only moves the spacecraft from an initial attitude angle to a final angle, but also reduces the elastic vibration caused by the maneuver. This chapter first gives the design method of the attitude maneuver path for flexible spacecraft with constant parameters based on input shaping. Then the variable amplitudes input shaping (VAIS) method and the procedure for optimization of controlled torque are presented. Focused on the slew maneuver of flexible spacecraft with parametric uncertainties, this chapter designs robust variable amplitudes input shapers including the variable amplitudes zero vibration derivative (ZVD) input shaper and the zero-vibration double-derivative (ZVDD) input shaper. And the corresponding robust attitude maneuver strategy is proposed.

5.2 Control Method of Input Shaping Attitude of Fixed Parameter System

In this section, for the flexible spacecraft with constant parameters, the design method of the maneuver path according to the input shaping method is studied, and the effectiveness of the method is verified by numerical simulation.

5.2.1 *Design of Attitude Maneuver Strategy Based on Input Shaping*

According to the Lagrange's equations in terms of quasi-coordinates, the mathematical model of the flexible spacecraft can be expressed as

$$\begin{aligned}&\boldsymbol{J}\dot{\boldsymbol{\omega}} + \boldsymbol{D}\ddot{\boldsymbol{\eta}} + \boldsymbol{\omega} \times \left(\boldsymbol{J}\boldsymbol{\omega} + \boldsymbol{D}\dot{\boldsymbol{\eta}}\right) = \boldsymbol{\tau}\\&\boldsymbol{D}^T\dot{\boldsymbol{\omega}} + \ddot{\boldsymbol{\eta}} + 2\boldsymbol{\xi}\boldsymbol{\Lambda}\dot{\boldsymbol{\eta}} + \boldsymbol{\Lambda}^2\boldsymbol{\eta} = \boldsymbol{0}\end{aligned} \tag{5.1}$$

For simplification, let us consider a rest-to-rest slewing maneuver of the spacecraft with rotating flexible appendage, as shown in Fig. 5.1. At time $t = 0$, the spacecraft is at rest with attitude angles $\boldsymbol{Q}_0$. Then the system undergoes rotation, and at the time t_{end}, the spacecraft should reach the desired final attitude angles $\boldsymbol{Q}_{end}$. Moreover, at the beginning and the end, angular velocities of the system are both zeroes.

For a rigid spacecraft, the maneuver time from a current state to an objective state with the bang-bang style is assumed as $2t_a$. The acceleration time and deceleration time are both t_a. At time $t = t_a$, the positive bang changes to the negative bang. The angular acceleration can be expressed by

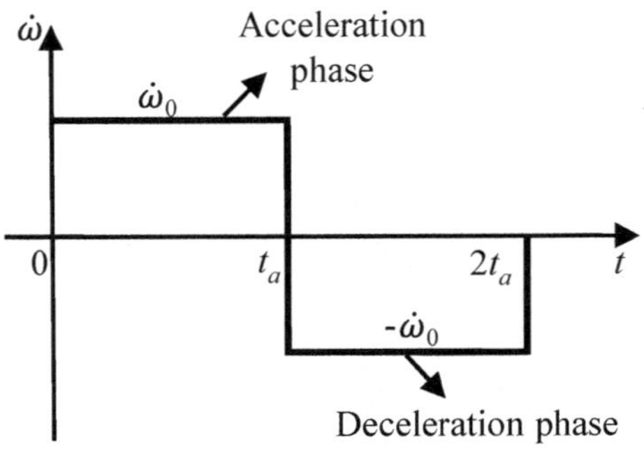

Fig. 5.1 Bang-bang profile example

$$\dot{\omega}(t) = \begin{cases} \dot{\omega}_0 & 0 < t < t_a \\ -\dot{\omega}_0 & t_a < t < 2t_a \end{cases} \tag{5.2}$$

A simple uncoupled linear model is used for designing slew trajectories [41]

$$\boldsymbol{J}(t)\dot{\boldsymbol{\omega}} = \boldsymbol{\tau}(t) \tag{5.3}$$

When the input shaper is used, the control input after shaping is

$$\boldsymbol{\tau}_s = \sum_{i=0}^{n} A_i \delta(t - t_i) * \boldsymbol{\tau}_0 \tag{5.4}$$

where $\sum_{i=0}^{n} A_i \delta(t - t_i)$ is the pulse sequence of the traditional input shaper, A_i is the pulse amplitude, t_i is the location time of the pulse, and the symbol $*$ represents convolution. The types of shapers include zero vibration (ZV), zero vibration and derivative (ZVD), and zero vibration and double derivative (ZVDD) [42].

The reference path is

$$\begin{aligned} \ddot{\boldsymbol{\theta}}_{ref}(t) &= \boldsymbol{J}^{-1}\boldsymbol{\tau}_s \\ \dot{\boldsymbol{\theta}}_{ref}(t) &= \dot{\boldsymbol{\theta}}_0 + \boldsymbol{J}^{-1}\boldsymbol{\tau}_s t \\ \boldsymbol{\theta}_{ref}(t) &= \boldsymbol{\theta}_0 + \dot{\boldsymbol{\theta}}_0 t + \boldsymbol{J}^{-1}\boldsymbol{\tau}_s \frac{t^2}{2} \end{aligned} \tag{5.5}$$

Transform the reference path in the inertial coordinate into the body coordinate,

$$\begin{aligned} \boldsymbol{\omega}_{ref}(t) &= \boldsymbol{T}(\boldsymbol{\theta}_{ref})\dot{\boldsymbol{\theta}}_{ref} \\ \dot{\boldsymbol{\omega}}_{ref}(t) &= \frac{d}{dt}\left[\boldsymbol{T}(\boldsymbol{\theta}_{ref})\right]\dot{\boldsymbol{\theta}}_{ref} + \boldsymbol{T}(\boldsymbol{\theta}_{ref})\ddot{\boldsymbol{\theta}}_{ref} \end{aligned} \tag{5.6}$$

where $\boldsymbol{T}(\boldsymbol{\theta}_{ref})$ is the transformation matrix from the angular rate vector $\dot{\boldsymbol{\theta}}_{ref}$ in the inertial coordinate to the angular rate $\boldsymbol{\omega}_{ref}$ in the body coordinate

$$\boldsymbol{T}(\boldsymbol{\theta}_{ref}) = \begin{bmatrix} c\theta & 0 & -s\theta \\ 0 & 1 & 0 \\ s\theta & 0 & c\theta \end{bmatrix} \begin{bmatrix} 1 & 0 & 0 \\ 0 & c\varphi & s\varphi \\ 0 & -s\varphi & c\varphi \end{bmatrix} \begin{bmatrix} c\psi & s\psi & 0 \\ -s\psi & c\psi & 0 \\ 0 & 0 & 1 \end{bmatrix} \tag{5.7}$$

The controlled torque for the spacecraft model can be determined as

$$\boldsymbol{\tau}_{ref}(t) = \begin{cases} \boldsymbol{J}\dot{\boldsymbol{\omega}}_{ref} + \boldsymbol{\omega}_{ref} \times \boldsymbol{J}\boldsymbol{\omega}_{ref} & (1) \\ \boldsymbol{J}\dot{\boldsymbol{\omega}}_{ref} + \boldsymbol{\omega}_{ref} \times \boldsymbol{J}\boldsymbol{\omega}_{ref} + \boldsymbol{D}\ddot{\boldsymbol{\eta}} & (2) \end{cases} \tag{5.8}$$

When using the control input represented by the above Eq. (5.1), the frequency of the system should be used in the input shaping calculation, and when using (2), the frequency of the flexible structure should be used.

5.2.2 Simulation of Attitude Maneuver Strategy Based on Input Shaping

In this section, the model of the flexible spacecraft system is used to perform three-axis attitude maneuver. The solar wing is fixed on the spacecraft platform, so the system parameters are constant during the simulation process.

The model for flexible spacecraft in Sect. 3.3 is utilized. At the initial moment, the Euler angle of the spacecraft is [0, 0, 0], the attitude angular velocity vector is [0, 0, 0], the target value of the Euler angle is [60, 45, 30]deg, and the maneuver time is 20 s. Both the ZV shaper and the ZVD shaper are used in simulation. When calculating the control input, the solar wing modes are used. Only the first three modes of the solar wing or the system are retained in the shaper.

(1) **Solar wing mode & ZV shaper**

Figure 5.2 shows the relative amplitude curve of angular acceleration when using ZV shaper, ZVD shaper and bang-bang maneuver. It can be seen that after using the shaper, the maneuvering time is extended.

Figures 5.3, 5.4, 5.5, 5.6 and 5.7 show the control input and dynamic responses of the system. Figures 5.4 and 5.5 are responses of the attitude angular velocity and Euler angle of the spacecraft platform respectively. It can be seen from the figure that with unshaped input, the attitude angular velocity will jitter due to the vibration of the solar wing after the attitude maneuver. The Euler angle reaches the target after the maneuver is completed.

Figures 5.6 and 5.7 are the modal response and tip displacement of the solar wing, respectively. During the attitude maneuver, low-order vibrations of the solar

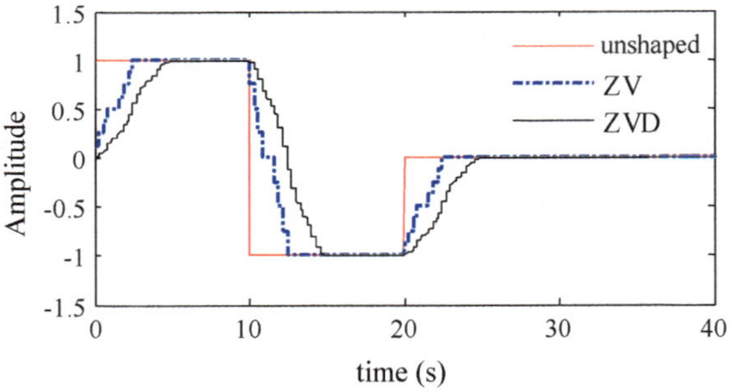

Fig. 5.2 Time series of relative amplitude of attitude angular acceleration

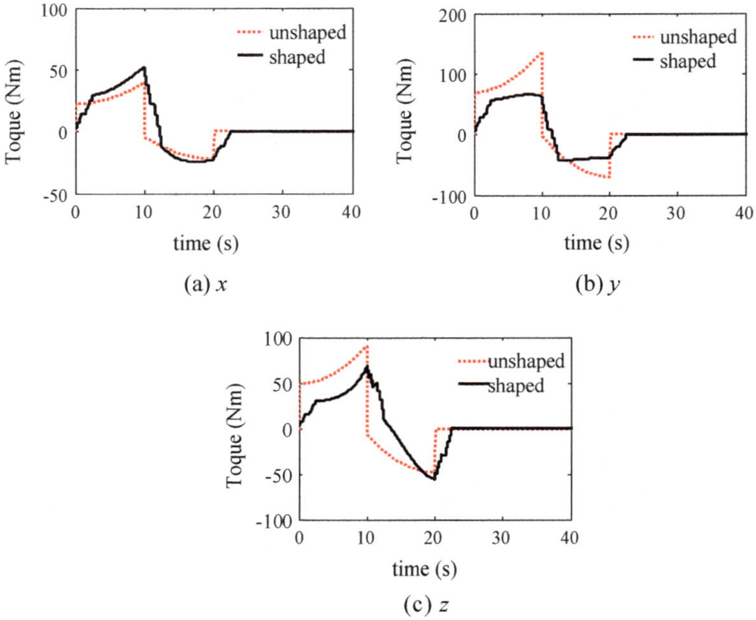

Fig. 5.3 Controlled torque

wing are excited. The tip displacement response mainly depends on the first-order mode. The y-direction displacement is dominated by the second-order mode (ie, the first-order in-plane bending mode), and the z-direction displacement is dominated by the first-order mode (ie, the first-order out-of-plane bending mode). The residual vibration of the solar wing is almost zero when the input shaper is applied.

(2) **Comparison of robustness between ZV and ZVD shapers**

In order to analyze the robustness of the attitude maneuvering method, the natural frequency of the solar wing is perturbed. Figure 5.8 shows the modal response of

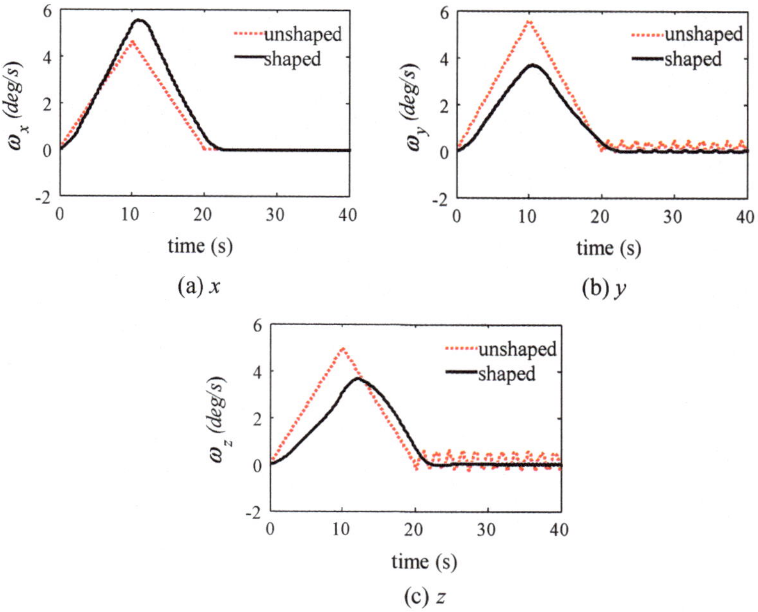

Fig. 5.4 Angular velocity response of the spacecraft

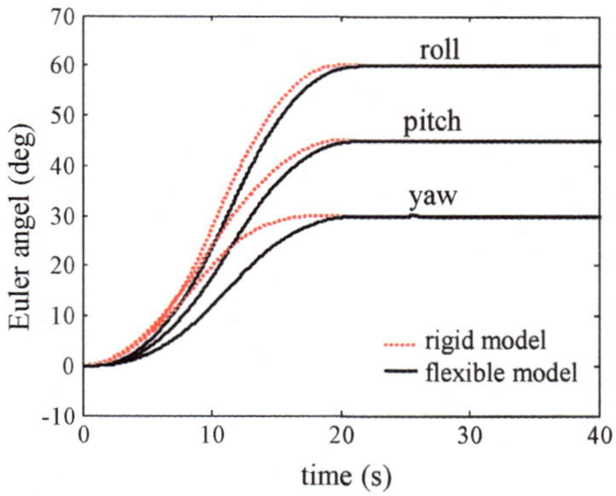

Fig. 5.5 Euler angle response of spacecraft

the solar wing when there exists an error of 5% in the natural frequency of the solar wing. Compared with the ZV shaper, the ZVD shaper is more robust.

The following conclusions can be drawn from this section:

(1) When the spacecraft with constant parameters is performing attitude maneuvers, the input shaping method can effectively reduce or even eliminate the residual vibration of the flexible structure after the maneuver;

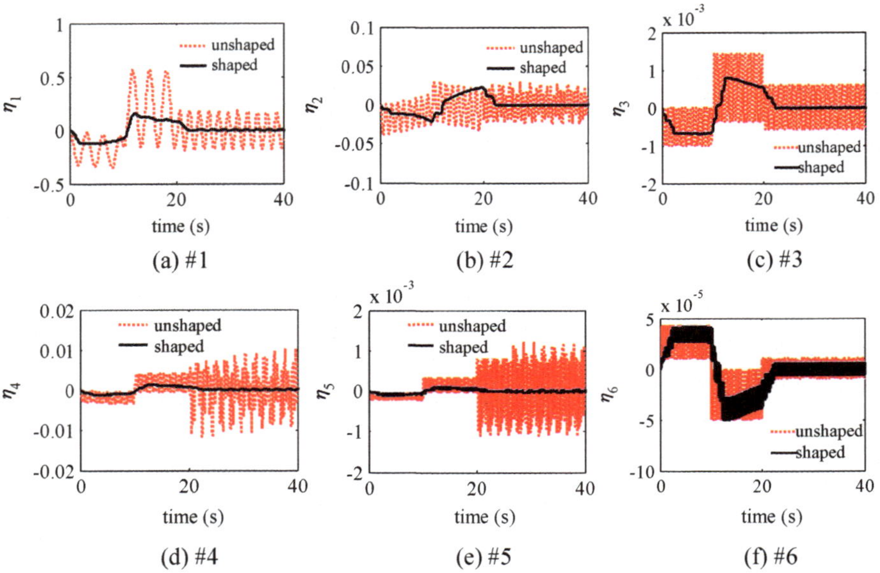

Fig. 5.6 Modal response of solar wing

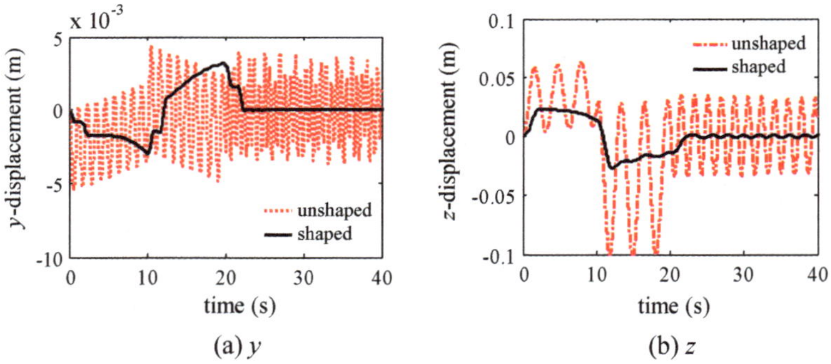

Fig. 5.7 Tip displacement response of solar wing

(2) The shaper can adopt the constrained modes of flexible structure or the system modes of spacecraft. However, when using the constrained modes of flexible structure, the natural frequencies and damping require to be accurate enough.

(3) When the natural frequency is deviated, the effect of the shaper will decrease. Compared with the ZV shaper, the ZVD shaper is more robust and is less affected by the modeling errors.

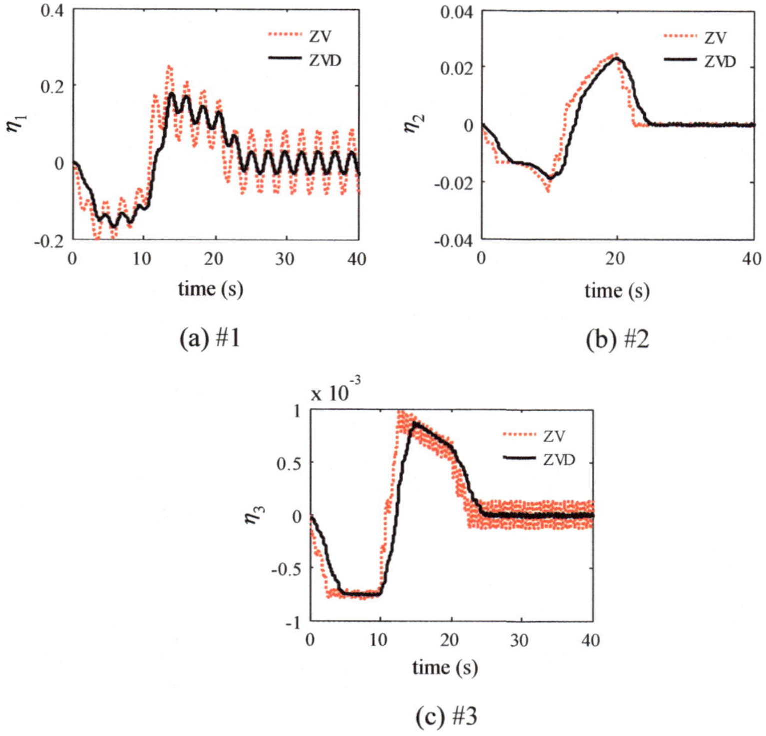

Fig. 5.8 Modal response of solar wing under two shapers at 5% frequency perturbation

5.3 Optimal Variable Amplitudes Input Shaping Control for Slew Maneuver[1]

In this section, a new design method for three-dimensional maneuver trajectory is developed for flexible spacecraft with time-varying parameters. The objective of the proposed theory is to maneuver the spacecraft to the desired attitude state, at the same time, cancel the residual vibration of the rotating flexible appendages after the maneuver. Unlike the classical input shaper method, this approach considers the effects of the time-varying parameters in the spacecraft mathematic model. The control input is discretized as a number of pulses. Linear equality constraints on pulses are set to guarantee the appendages' residual vibration to be zero based on the input shaper idea with time-varying impulse sequence and nonlinear constraints are set to drive the main platform to the desired state. Then the control input is optimized to ensure smoothness.

[1] Jie Wang, Dongxu Li, Jianping Jiang. Optimal variable amplitudes input shaping control for slew maneuver of flexible spacecraft, Journal of Guidance, Control, and Dynamics, 2017, 12, 3255–3263.

5.3.1 Optimal Variable Amplitudes Input Shaping Control

The procedure for optimal slew trajectory design is depicted in Fig. 5.9. First, after given the slew task, the system's rigid model is adopted to design the baseline trajectory. Second, based on the baseline trajectory, an optimization problem is defined to derive the optimal trajectory for flexible model. During the establishment of the optimization problem, the variable amplitudes input shaping is proposed for systems with time-varying parameters.

5.3.1.1 Baseline Trajectory Design

As the first step, a simple uncoupled linear model is used for designing slew trajectories. Specify the angular acceleration vector as

$$\boldsymbol{a} = \dot{\boldsymbol{\omega}} \tag{5.9}$$

It can be noted that the acceleration vector has a similar profile to the controlled torque. In the acceleration phase, the acceleration vector is $\boldsymbol{a}_0 = [a_{x,0}, a_{y,0}, a_{z,0}]^{\mathrm{T}}$ while the vector in deceleration phase is -$\boldsymbol{a}_0$. The baseline controlled torque for the spacecraft model can be determined as

$$\boldsymbol{\tau}_{ref}(t) = \boldsymbol{J}(t)\boldsymbol{a} + \boldsymbol{\omega} \times \boldsymbol{J}(t)\boldsymbol{\omega} + \boldsymbol{D}(t)\ddot{\boldsymbol{\eta}} \tag{5.10}$$

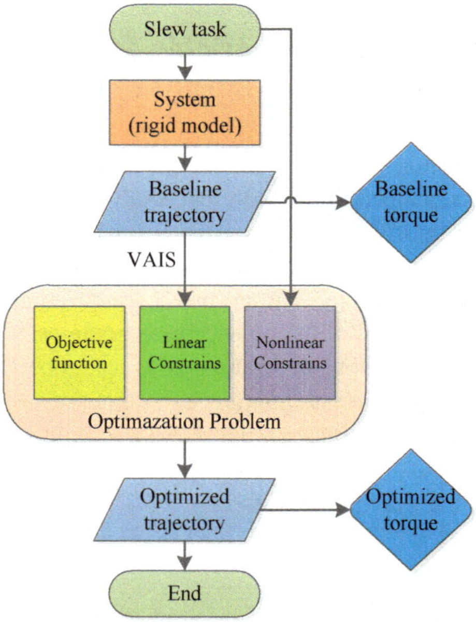

Fig. 5.9 Flowchart of optimal slew trajectory design for flexible spacecraft

5.3.1.2 Variable Amplitudes Input Shaping

The basic idea of ZV (zero vibration) input shaper is shown in Fig. 5.10. Given a second-order generalized linear system in the form

$$\begin{cases} \ddot{x} + 2\varsigma\omega\dot{x} + \omega^2 x = F(t) \\ x(0) = 0, \qquad \dot{x}(0) = 0 \end{cases} \tag{5.11}$$

where x is the state coordinate of the system, ς is damping ratio, ω is the circular frequency, and $F(t)$ is the external excitation. ω_d represents the damped vibration frequency, $\omega_d = \omega\sqrt{1-\varsigma^2}$. T denotes the corresponding period, $T = 2\pi/\omega_d$.

The shaper is composed of two pulse signals. The pulse signal is convolved with the reference signal to generate the final shaped signal. In addition, the amplitude of the two pulse signals must meet a certain relationship, so that the two signal components after forming produce the same rigid body motion and minimize the residual vibration. The time history of the reference signal is represented by the symbol t_f, and it is assumed that the signal is stable at time $t = t_r$ ($0 \le t_r \le t_f$), which is a constant value. The time history of the shaper is half the period of the damping system, so the time history of the signal after molding is $t_f + T/2$.

When amplitudes of the two pulse signals yield the following relationship

$$\text{ZV} = \begin{bmatrix} A_i \\ t_{A_i} \end{bmatrix} \begin{bmatrix} \frac{1}{1+K} & \frac{K}{1+K} \\ 0 & \frac{T}{2} \end{bmatrix} \quad (i = 1, 2) \tag{5.12}$$

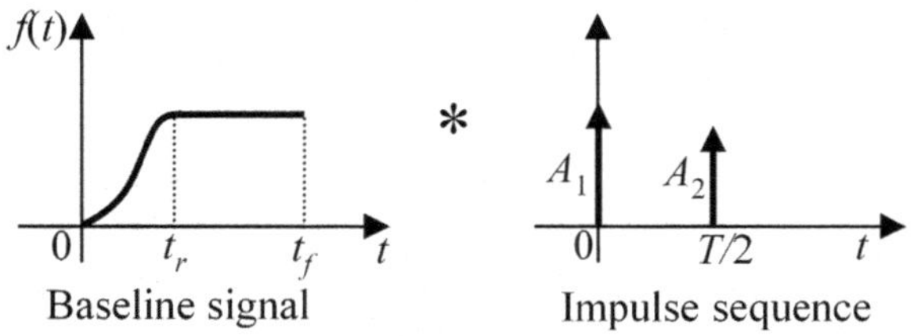

Fig. 5.10 Schematic of ZV input shaper

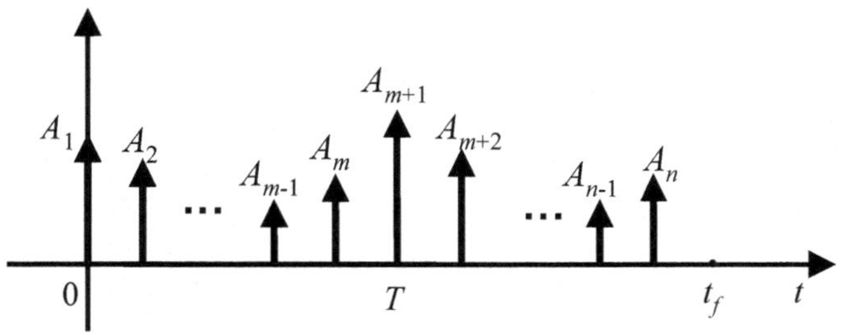

Fig. 5.11 Time-varying impulse sequence

where

$$K = e^{-\varsigma\pi/\sqrt{1-\varsigma^2}} \tag{5.13}$$

The force of the following form

$$\begin{aligned} F(t) &= f(t) * [A_1\delta(t)] + f(t) * [A_2\delta(t - T/2)] \\ &= A_1 f(t) * \delta(t) + A_2 f(t) * \delta(t - T/2) \end{aligned} \tag{5.14}$$

will make the vibration of the system zero at time $t > t_a + T/2$. Where $\delta(t)$ is the Dirac function.

Substituting Eq. (5.9) into Eq. (2.135) leads to

$$\ddot{\boldsymbol{\eta}} + 2\boldsymbol{\xi}\boldsymbol{\Lambda}\dot{\boldsymbol{\eta}} + \boldsymbol{\Lambda}\boldsymbol{\eta} = -\boldsymbol{D}(t)^T\boldsymbol{a} \tag{5.15}$$

The system formulated in the above equation is a second-order generalized linear system on which the standard command input shaping method can be used to design the external input. Also, because of the existence of a time-varying term, a modified input shaping technique with time-varying amplitudes sequence is proposed to accommodate the new system.

We propose a modified input shaping technique with time-varying amplitudes sequence based on the idea of ZV shaper. Divide the duration t_f into n ($n \in \mathbb{Z}^+$) identical components of t_f/n interval, which is denoted as Δt. Also we assume that the period T yields $T = m \cdot (t_f/n)$, ($m \in \mathbb{Z}^+$) (Fig. 5.11).

The control input can be decomposed into a number of sequences

$$\begin{aligned} F(t) &= \sum_{i=1}^{n} f_i(t) * \delta(t - t_i) \\ &= f_1(t) * \delta(t - t_1) + f_2(t) * \delta(t - t_2) + \cdots \\ &\quad + f_m(t) * \delta(t - t_m) + f_{m+1}(t) * \delta(t - t_{m+1}) + \cdots + f_n(t) * \delta(t - t_n) \end{aligned} \tag{5.16}$$

where $f_i(t)$ is the i-th input component, t_i denotes the delay time, and $t_i = (i-1) \cdot \Delta t$.

According to the principal of ZV, control input in Eq. (5.16) leads to suppression of the vibration mode in Eq. (5.15) if the input component satisfies

$$\begin{cases} \sum\limits_{k=1}^{j+(k-1)m \le n} e^{-\varsigma\omega_d t_{j+(k-1)m}} f_{j+(k-1)m}(t) = 0 & (j = 1, 2, \ldots, m) \quad if\ m\ is\ odd \\ \sum\limits_{k=1}^{j+(k-1)\frac{m}{2} \le n} (-1)^{k-1} e^{-\varsigma\omega_d t_{j+(k-1)m}} f_{j+(k-1)\frac{m}{2}}(t) = 0 & \left(j = 1, 2, \ldots, \frac{m}{2}\right) \quad if\ m\ is\ even \end{cases} \tag{5.17}$$

It should be noted that when the input components are all functions with constant values, namely $t_r = 0$, the system will reach a steady state with zero vibration after the time $t = t_f$.

5.3.1.3 Slew Maneuver Trajectory Optimization

Now, we design the slew trajectory for time-varying parameters system according to input shaper with time-varying impulse sequence. The angular acceleration vector is transformed to a dimensionless form

$$\boldsymbol{a} = \boldsymbol{\chi}\boldsymbol{a}_0 = \begin{bmatrix} \chi_x(t) & & \\ & \chi_y(t) & \\ & & \chi_z(t) \end{bmatrix} \begin{bmatrix} a_{x,0} \\ a_{y,0} \\ a_{z,0} \end{bmatrix} \tag{5.18}$$

where χ_x, χ_y, χ_z are ratios in dimensionless form. And for the bang-bang input, $\boldsymbol{\chi}$ satisfies

$$\chi_i = \begin{cases} 1 & 0 < t < t_a \\ -1 & t_a < t < 2t_a \end{cases} \quad (i = x, y, z) \tag{5.19}$$

The time-varying dynamic model is simplified as a piecewise linear model. The duration t_a is divided into n parts and the interval is denoted as Δt. In each interval, the inertia matrix and coupling matrix are both assumed as constant matrices.

Refer to the bang-bang control, the control input profile for the piecewise time-invariable system is illustrated in Fig. 5.12. Compared to the time-invariable system, the accelerating time remains t_a. In each interval, the control input remains as constant. The first N periods corresponding to the first N natural frequencies for the flexible appendage are denotes as $T_1, T_2, \ldots T_N$ $(T_1 > T_2 > \ldots > T_N)$, and meet the following conditions

$$m_1 \Delta t = T_1$$

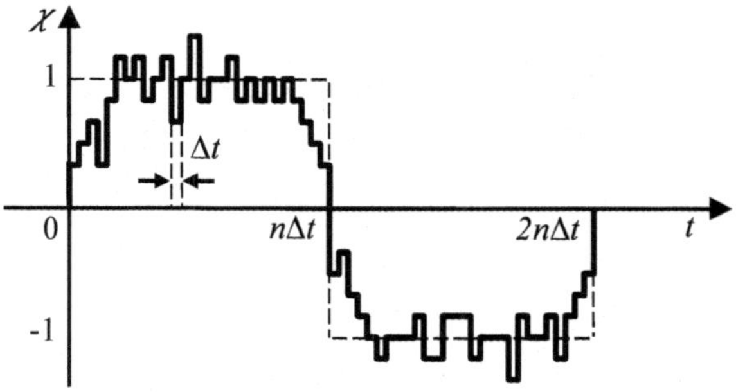

Fig. 5.12 Control profile of bang-bang maneuver for time-varying system

$$m_2 \Delta t = T_2$$
$$\vdots$$
$$m_N \Delta t = T_N \tag{5.20}$$

In each interval, the external force applied to the flexible system remains constant. For instance, in the i-th $(1,2,\ldots,n)$ interval $(t \in [(i-1)\Delta t \; i\Delta t])$, the force has the form

$$\boldsymbol{F}_i = -\boldsymbol{D}_i^T \boldsymbol{a}_i \tag{5.21}$$

The force can be decomposed as

$$\begin{aligned} \boldsymbol{F}(t) = & \boldsymbol{F}_1 * \delta(t - t_1) + (\boldsymbol{F}_2 - \boldsymbol{F}_1) * \delta(t - t_2) + \cdots \\ & + (\boldsymbol{F}_{\mathrm{m}} - \boldsymbol{F}_{\mathrm{m-1}}) * \delta(t - t_m) + (\boldsymbol{F}_{\mathrm{m+1}} - \boldsymbol{F}_{\mathrm{m}}) * \delta(t - t_{m+1}) + \cdots \\ & + (\boldsymbol{F}_n - \boldsymbol{F}_{\mathrm{n-1}}) * \delta(t - t_n) \end{aligned} \tag{5.22}$$

Define

$$\boldsymbol{f}_{i+1} = \begin{bmatrix} f_{1,i+1} & f_{1,i+1} & \cdots & f_{N,i+1} \end{bmatrix}^T = \boldsymbol{F}_{i+1} - \boldsymbol{F}_i \tag{5.23}$$

According to Eq. (5.17), the following conditions are proposed in order to reduce the residual vibration of the flexible appendage after the slew maneuver

$$\begin{aligned} & for\ mode\ \#1: && \sum_{j=1}^{k+(j-1)m_1 \le n} e^{-\xi_1 \Lambda_1 [k+(j-1)m_1]\Delta t} f_{1,k+(j-1)m_1} = 0 && (k = 1, 2, \ldots, m_1) \\ & for\ mode\ \#2: && \sum_{j=1}^{k+(j-1)m_2 \le n} e^{-\xi_2 \Lambda_2 [k+(j-1)m_2]\Delta t} f_{2,k+(j-1)m_2} = 0 && (k = 1, 2, \ldots, m_2) \\ & \vdots && \vdots \qquad\qquad \vdots && \\ & for\ mode\ \#N: && \sum_{j=1}^{k+(j-1)m_N \le n} e^{-\xi_N \Lambda_N [k+(j-1)m_N]\Delta t} f_{N,k+(j-1)m_N} = 0 && (k = 1, 2, \ldots, m_N) \end{aligned} \tag{5.24}$$

The above constraints are set to make the elastic vibration converge to zero after the maneuver. The number of equality constraint functions in the above equation equals to $\sum_{p=1}^{N} m_p$.

Besides, the shaped input should guarantee the attitude angles to be consistent with the bang-bang maneuver. The equation in quaternion form can be discretized as

$$\dot{\boldsymbol{Q}}_i = \frac{1}{2}\Omega(\boldsymbol{\omega}_i)\boldsymbol{Q}_i \tag{5.25}$$

So the quaternion vector $\boldsymbol{Q}_{i+1}$ at time $t = i\Delta t$ is

$$\boldsymbol{Q}_{i+1} = \left(\boldsymbol{I}_4 + \frac{1}{2}\Omega(\boldsymbol{\omega}_i)\Delta t\right)\boldsymbol{Q}_i \tag{5.26}$$

Define the state transition matrix from quaternion vector at time $t = i\Delta t$ to the vector at time $t = (i+1)\Delta t$ as

$$\boldsymbol{T}_{i-1\to i} = \boldsymbol{I}_4 + \frac{1}{2}\Omega(\boldsymbol{\omega}_i)\Delta t \tag{5.27}$$

So the state transition matrix from the final quaternion vector and initial vector can be derived as follows

$$\mathbf{T}_{0\to n} = \prod_{i=1}^{n}\left(\boldsymbol{I}_4 + \frac{1}{2}\Omega(\boldsymbol{\omega}_i)\Delta t\right) \tag{5.28}$$

Therefore, the state transition matrix of the duration should yields

$$\boldsymbol{Q}_{end} = \mathbf{T}_{0\to n}\boldsymbol{Q}_0 \tag{5.29}$$

The above constraints were derived from the requirements that the amplitude of the Euler angles be made identical with the case under the bang-bang maneuver. It should be noted that there are four equality constraint functions in roll, pitch and yaw directions.

To reduce the sudden change in control input to the system, a smoothing function υ is defined as follows

$$\upsilon(\chi) = \sqrt{\sum_{i=1}^{n}\left(\chi_i - \frac{1}{n}\sum_{i=1}^{n}\chi_i\right)^2} \tag{5.30}$$

Thus, the problem is equivalent to find a $\boldsymbol{\chi}$ under the constraints (5.24) and (5.29). An optimization problem can be described as follows

$$\begin{aligned} &fing\ \chi \\ &\min\ \upsilon(\chi) \\ &\text{s.t}\ \begin{cases} Eq.\,(5.24) \\ Eq.\,(5.29) \end{cases} \end{aligned} \tag{5.31}$$

Because the constraints in Eq. (5.29) is nonlinear, the SQP (sequential quadratic programming) algorithm is applied to obtain globally optimal solutions. With the optimal $\boldsymbol{\chi}$, the optimized slew trajectory and the corresponding controlled torque can be calculated by Eqs. (5.18) and (5.10) respectively.

5.3.2 Numerical Simulation

This section verifies the effectiveness of the method proposed in the previous section through a simulation example. The simulation model uses the model proposed in Sect. 3.3.1. For simplicity, we only consider the first six cantilevered flexible modes, including the first three y_s-axial bending mode, the first z_s-axial bending mode, and the first two x_s-axial torsional modes. Natural frequencies of the six modes are 0.32 Hz, 0.88 Hz, 1.79 Hz, 1.80 Hz, 4.51 Hz, and 5.19 Hz respectively. And corresponding periods are 3.16 s, 1.14 s, 0.56 s, 0.55 s, 0.22 s, and 0.19 s. Also, the first three modes are considered during designing the input shaper.

Consider a three-dimensional rest-to-rest maneuver case. Initially the spacecraft is balanced at an initial Euler angle $[0, 0, 0]^{\mathrm{T}}$ with initial angular velocity $\boldsymbol{\omega}(0) = [0, 0, 0]^{\mathrm{T}}$. And the desired Euler angle is $[60°, 45°, 30°]^{\mathrm{T}}$. The span of the slew maneuver is 20 s. Solar arrays are hinged to the platform and rotates about the x_s axis with a constant angular velocity $\Omega = 3\ deg/s$. At time $t = 0$, the parameter γ equals to 0°.

At time $t = 0$, an optimized torque vector applies to the platform. At time $t \geq 20\ s$, controlled torques around the three axes are all set to zero. The controlled torques for the rigid model in which the flexibility of appendages is neglected are calculated by Eq. (5.10). And the shaped controlled torques are derived by optimizing the problem defined in Eq. (5.31).

Figure 5.13 shows the control inputs on the roll, yaw and pitch degrees-of-freedom for the rigid model (in a dotted line) and flexible model (in a solid line). With the input shaper considered, the control input is optimized to be a combination of a number of pulses. The control input is designed to suppress up to the first three vibration modes.

Figure 5.14 shows the angular velocity to the unshaped and shaped controlled torque. Under the control input derived from the rigid model, the angular rates experience severe oscillations at the end of the control input because of the solar arrays' vibration. With the first three modes considered, the shaped control input induces very little vibration for the angular rates at the end of the control input.

Figure 5.15 gives the tip displacement of the solar array in the floating coordinate. As expected, the shaped input activates much lower magnitude than the unshaped input. The magnitudes of elastic vibration of the three modes are ultimately brought to near zero. Consequently, after the controlled torque is removed, the terminal oscillations of Euler's angles have much lower magnitudes. Also, displacements of the tip along y-direction and z-direction are dominated by the first two modes (Fig. 5.16). The value of the fourth natural frequency is close to the third one. Thus, the vibration of the fourth mode is reduced under the shaped input. The vibrations of the fifth and sixth modes are not reduced because they are not considered in the input shaper.

Figure 5.17 shows the Euler angle response of the spacecraft body under the controlled torque before and after forming. It can be seen that under the action of the controlled torque before forming, the Euler angle oscillates and there is an offset, which is due to the posture angular velocity oscillation after the maneuver is completed and the equilibrium position is not zero. Under the controlled torque after

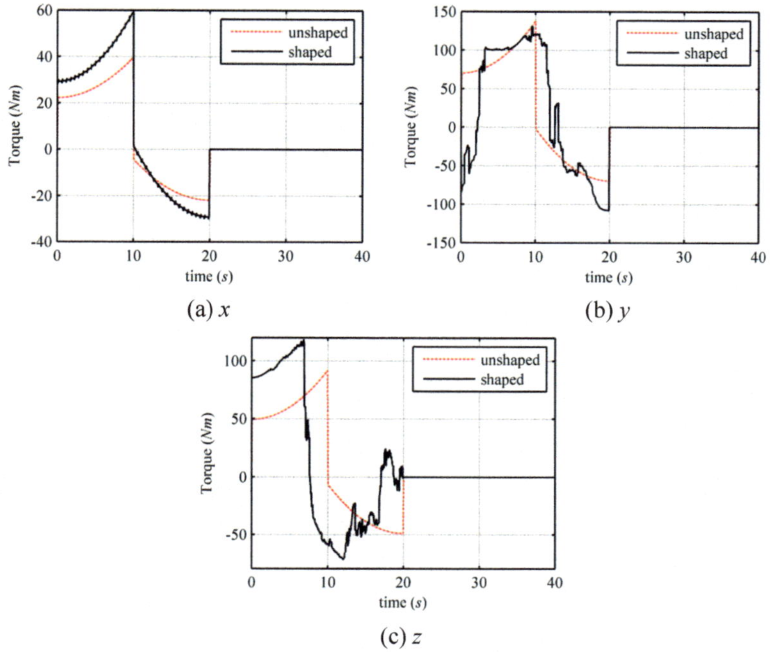

Fig. 5.13 Controlled torque of the flexible spacecraft

forming, the Euler angle reaches the target value after the maneuver is completed, and there is no oscillation (Fig. 5.18).

Simulation results show that the proposed control method is effective and achieves the control goal of reaching the target value of the attitude angle and zero residual vibration for the flexible structure.

To evaluate the robustness to modeling errors and parameter variations, a perturbation of natural frequencies of the solar wing is given and, in this situation, transient responses of the system are derived. Define a difference factor of natural frequencies as $\varepsilon = |1-(\omega_{act}/\omega_{nom})|$, where ω_{act} is the actual frequency of the solar array, and ω_{nom} is the nominal frequency which is used to design the shaper. Figure 5.19 illustrates the oscillation of the first three modes. It can be shown that the amplitudes of residual vibration increase prominently with the increase of the difference factor. Therefore, the input shaper applied is sensitive to model errors or other uncertainties in modeling.

In practical problems, it is difficult to accurately obtain the damping information of the system, and there are more or less errors in the estimated system damping. Therefore, the robustness of the shaper to system damping errors needs to be considered. When designing the attitude maneuvering path, the modal damping ratio of the flexible solar wing is assumed to be zero. The robustness of the attitude path to system uncertainty when the actual system damping ratio is 0.01 is investigated. Figure 5.20 shows the modal coordinate response of the solar wing when the actual

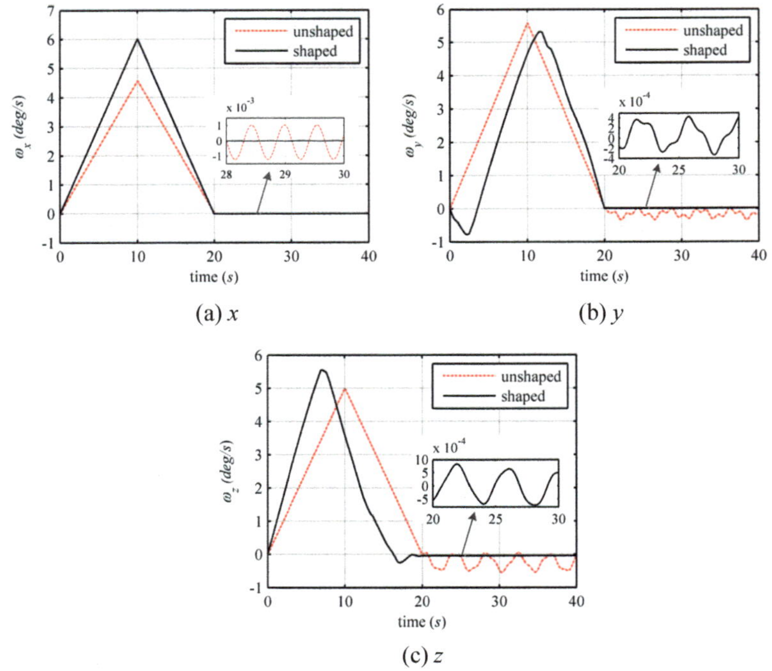

Fig. 5.14 Angular velocities of the flexible spacecraft

damping ratio is 0 and 0.01. At this time, the modal damping ratio errors are 0 and 0.01 respectively. It can be seen that when the variable amplitude ZV shaper is used, the residual vibration of the solar wing increases significantly when there is damping uncertainty. For the first-order modal coordinates, the residual vibration amplitude is about 0.02.

5.4 Robust Attitude Maneuver Strategy Based on Variable Amplitudes Input Shaping Method

In the previous section, a variable-amplitude zero-order input shaper was proposed for flexible spacecraft with variable parameters. However, the applied variable amplitudes zero vibration (ZV) input shaper is sensitive to modeling errors or other uncertainties in modeling. For this reason, robust variable amplitudes input shapers are designed in this paper, and a robust attitude maneuver strategy based on the shapers is proposed.

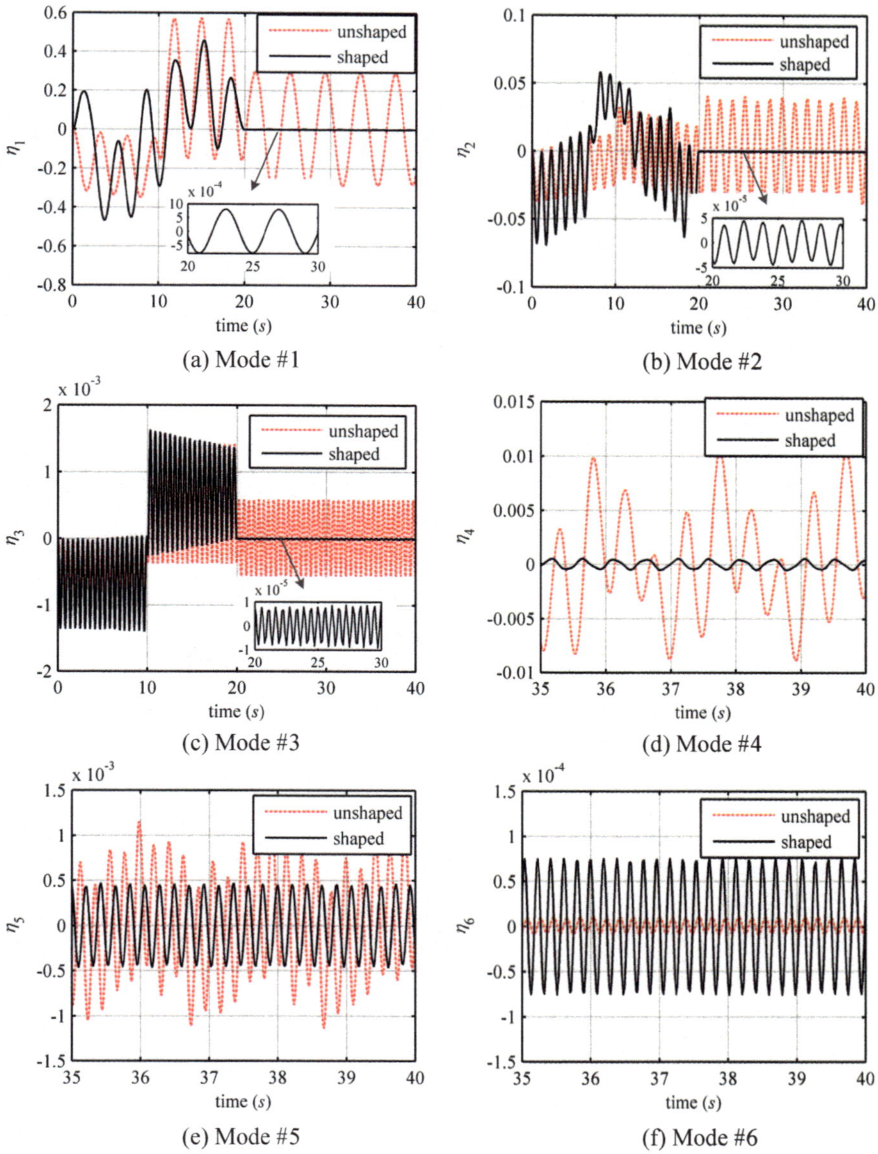

(a) Mode #1 (b) Mode #2

(c) Mode #3 (d) Mode #4

(e) Mode #5 (f) Mode #6

Fig. 5.15 Modal coordinates of the solar array

5.4.1 Robust Attitude Maneuver Strategy

Depending on literature [42], sensitivities of several input shapers including ZV, ZVD, and ZVDD, on frequency perturbation were researched. It can be noted that, with increasing orders of the input shaper, it is more robust to variations of system

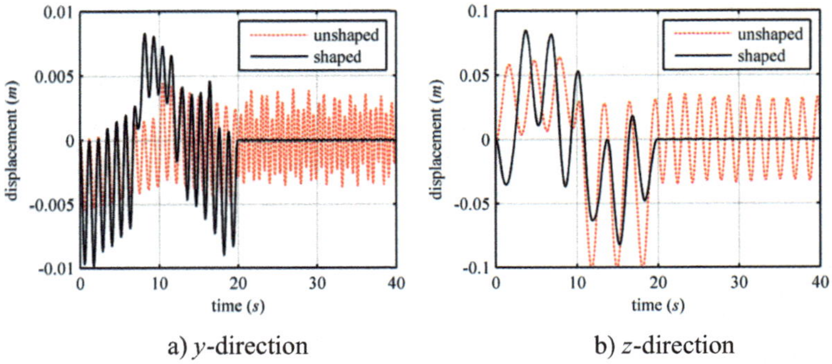

Fig. 5.16 Tip displacement of the solar array in floating coordinate

Fig. 5.17 Attitude of the flexible spacecraft

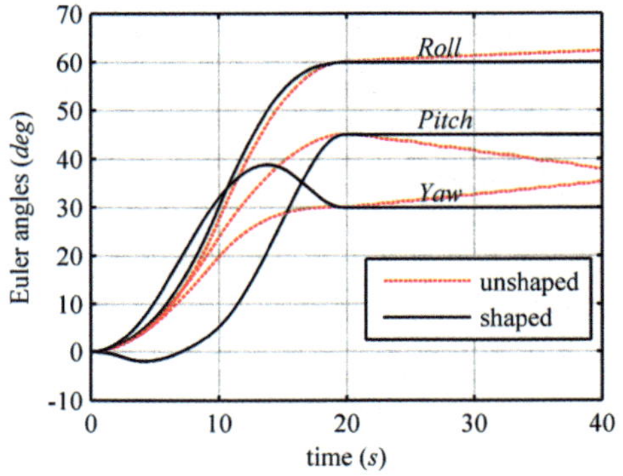

Fig. 5.18 Comparison the attitude angle

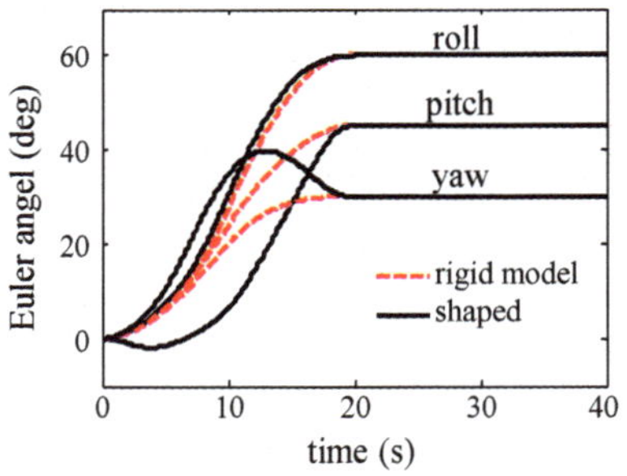

parameters. Hence, based on the variable amplitudes ZV shaper in this section we develop variable amplitudes ZVD and ZVDD shapers with higher orders in order to improve robustness of the VAIS.

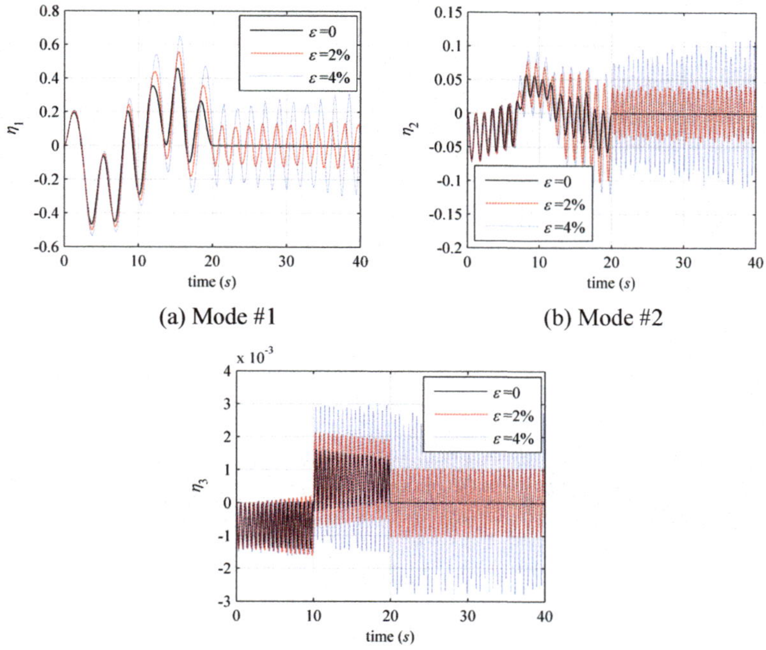

(a) Mode #1

(b) Mode #2

(c) Mode #3

Fig. 5.19 Responses of modal coordinates with frequency perturbation

5.4.1.1 Variable Amplitudes ZVD Input Shaper

Figure 5.21 presents the principle of ZVD input shaper, which is composed of three impulses. Time interval between each two pulses is half of period. The final signal is calculated by convolving the baseline signal convolves with the impulse sequence. Hence, the duration of the shaped signal is increased to be $t_f + T$.

In order to cancel vibration, amplitudes of the three impulses must satisfy

$$\text{ZVD} = \begin{bmatrix} A_i \\ \hline t_{A_i} \end{bmatrix} = \begin{bmatrix} \frac{1}{1+2K+K^2} & \frac{2K}{1+2K+K^2} & \frac{K^2}{1+2K+K^2} \\ 0 & \frac{T}{2} & T \end{bmatrix} \quad (i = 1, 2, 3) \tag{5.32}$$

Compared to the ZV input shaper, ZVD input shaper is more robust to parametric uncertainties. A variable amplitudes ZVD input shaper is then proposed based on the ZVD input shaper. Divide the duration t_f into n ($n \in \mathbb{Z}^+$) identical components of t_f/n interval, which is denoted as Δt. Also we assume that the period T yields $T = m \cdot (t_f/n)$, ($m \in \mathbb{Z}^+$).

The control input can be decomposed into a number of sequences

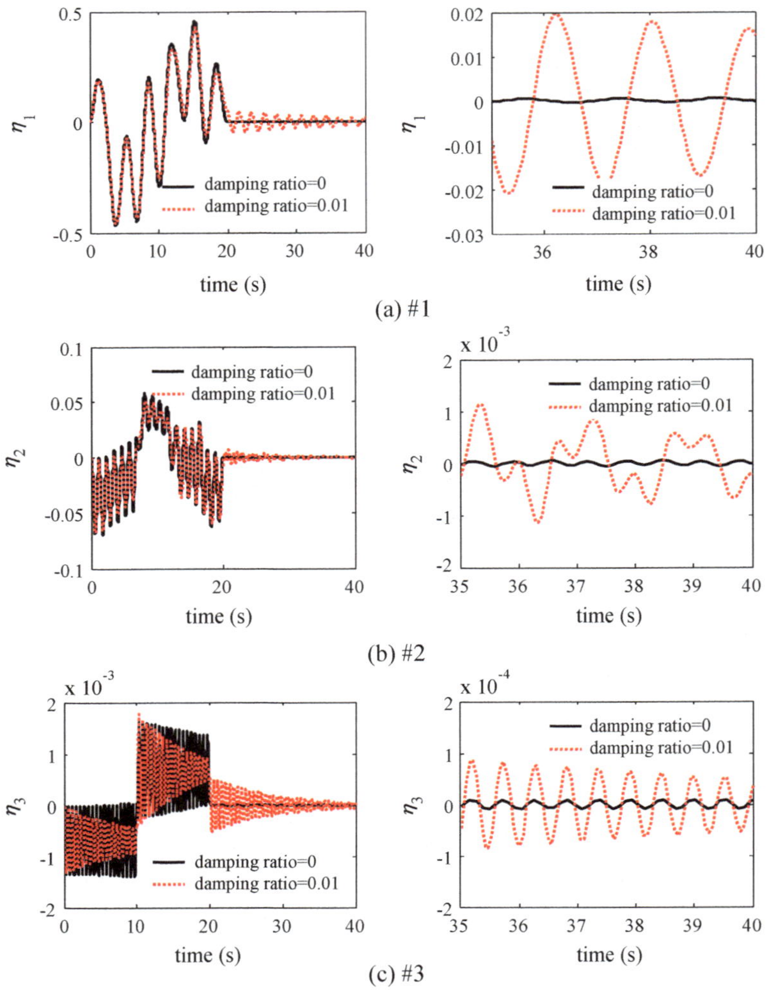

Fig. 5.20 Responses of modal coordinates with damping perturbation

Fig. 5.21 Schematic of ZVD input shaper

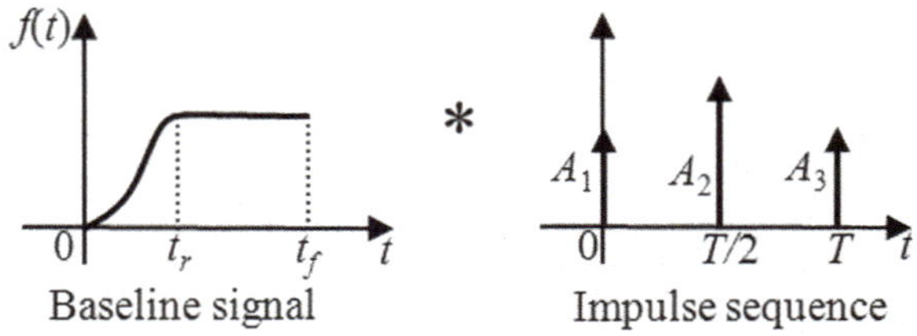

$$
\begin{aligned}
F(t) &= \sum_{i=1}^{n} f_i(t) * \delta(t - t_i) \\
&= f_1(t) * \delta(t - t_1) + f_2(t) * \delta(t - t_2) + \cdots \\
&\quad + f_m(t) * \delta(t - t_m) + f_{m+1}(t) * \delta(t - t_{m+1}) + \cdots + f_n(t) * \delta(t - t_n)
\end{aligned} \tag{5.33}
$$

where $f_i(t)$ is the i-th input component, t_i denotes the delay time, and $t_i = (i-1)\cdot\Delta t$.

Define a vector consisting of input components with intervals of integral multiple of half of the period

$$
\widehat{\boldsymbol{f}}_j = \left[f_j \; f_{j+m/2} \; f_{j+m} \; \cdots \; f_{j+(k-1)m/2} \right]^T \quad \left(\begin{array}{l} j = 1, 2, \ldots, \frac{m}{2} \\ j + (k-1)\frac{m}{2} \leq n \end{array} \right) \tag{5.34}
$$

Introduce components of $\widehat{\boldsymbol{f}}_j$ as

$$
\widehat{\boldsymbol{g}}_{j,l} = \left[g_{j,l} \; g_{j,l+1} \; g_{j,l+2} \right]^T \quad \left(\begin{array}{l} j = 1, 2, \ldots, m/2 \\ l = 1, 2, \ldots, k-2 \end{array} \right) \tag{5.35}
$$

Amplitudes of the scalars in $\widehat{g}_{j,l}$ satisfy constraints of ZVD, namely

$$
\frac{g_{j,l}}{1/\left(1 + 2K + K^2\right)} = \frac{g_{j,l+1}}{2K/\left(1 + 2K + K^2\right)} = \frac{g_{j,l+2}}{K^2/\left(1 + 2K + K^2\right)} \tag{5.36}
$$

Define $\overline{g}_{j,l}$ as the amplitude of the lth component. Hence, $\widehat{\boldsymbol{g}}_{j,l}$ has the expression of

$$
\widehat{\boldsymbol{g}}_{j,l} = \overline{g}_{j,l} \left[\frac{1}{1+2K+K^2} \; \frac{2K}{1+2K+K^2} \; \frac{K^2}{1+2K+K^2} \right]^T \tag{5.37}
$$

$\widehat{\boldsymbol{g}}_{j,l}$ can cancel residual vibration of the second-order system. The vector $\widehat{f}_j$ is seen as a linear combination of $\widehat{\boldsymbol{g}}_{j,l}$

$$
\begin{aligned}
\widehat{\boldsymbol{f}}_j(1) &= f_j = \widehat{\boldsymbol{g}}_{j,1}(1) \\
\widehat{\boldsymbol{f}}_j(2) &= f_{j+\frac{m}{2}} = \widehat{\boldsymbol{g}}_{j,2}(1) + \widehat{\boldsymbol{g}}_{j,1}(2) \\
&\vdots \\
\widehat{\boldsymbol{f}}_j(i) &= f_{j+(i-1)\frac{m}{2}} = \widehat{\boldsymbol{g}}_{j,i}(1) + \widehat{\boldsymbol{g}}_{j,i-1}(2) + \widehat{\boldsymbol{g}}_{j,i-2}(3) \quad (i = 3, 4, \ldots, k-2) \\
&\vdots \\
\widehat{\boldsymbol{f}}_j(k-1) &= f_{j+(k-2)\frac{m}{2}} = \widehat{\boldsymbol{g}}_{j,k-2}(2) + \widehat{\boldsymbol{g}}_{j,k-3}(3) \\
\widehat{\boldsymbol{f}}_j(k) &= f_{j+(k-1)\frac{m}{2}} = \widehat{\boldsymbol{g}}_{j,k-2}(3)
\end{aligned}
\tag{5.38}
$$

namely

$$
\widehat{\boldsymbol{f}}_j = \begin{bmatrix} f_j \\ f_{j+m/2} \\ f_{j+m} \\ \vdots \\ f_{j+(k-2)m/2} \\ f_{j+(k-1)m/2} \end{bmatrix} = \left[\begin{array}{ccc|ccc|ccc|c|ccc|ccc}
1 & & & & & & & & & & & & & & & \\
 & 1 & & 1 & & & & & & & & & & & & \\
 & & 1 & & 1 & & 1 & & & & & & & & & \\
 & & & & & 1 & & 1 & & & & & & & & \\
 & & & & & & & & 1 & & & & & & & \\
 & & & & & & & & & \ddots & & & & & & \\
 & & & & & & & & & & 1 & & & & & \\
 & & & & & & & & & & & 1 & & 1 & & \\
 & & & & & & & & & & & & 1 & & 1 & \\
 & & & & & & & & & & & & & & & 1
\end{array}\right] \begin{bmatrix} \widehat{\boldsymbol{g}}_{j,1} \\ \widehat{\boldsymbol{g}}_{j,2} \\ \widehat{\boldsymbol{g}}_{j,3} \\ \vdots \\ \widehat{\boldsymbol{g}}_{j,k-3} \\ \widehat{\boldsymbol{g}}_{j,k-2} \end{bmatrix}
\tag{5.39}
$$

Inserting Eq. (5.37) into Eq. (5.40), we can obtain

$$
\hat{f}_j = \begin{bmatrix} f_j \\ f_{j+m/2} \\ f_{j+m} \\ \vdots \\ f_{j+(k-2)m/2} \\ f_{j+(k-1)m/2} \end{bmatrix} = \begin{bmatrix}
A_1 & & & & & \\
A_2 & A_1 & & & & \\
A_3 & A_2 & A_1 & & & \\
 & A_3 & A_2 & & & \\
 & & A_3 & & & \\
 & & & \ddots & & \\
 & & & & A_1 & \\
 & & & & A_2 & A_1 \\
 & & & & A_3 & A_2 \\
 & & & & & A_3
\end{bmatrix} \begin{bmatrix} \bar{g}_{j,1} \\ \bar{g}_{j,2} \\ \bar{g}_{j,3} \\ \vdots \\ \bar{g}_{j,k-3} \\ \bar{g}_{j,k-2} \end{bmatrix}
\tag{5.40}
$$

The vector $\widehat{\boldsymbol{f}}_j$ in form of Eq. (5.33) can make residual vibration of the system to be zero. Therefore, it is only necessary to find a suitable one $\overline{g}_{j,l}$, to ensure that the

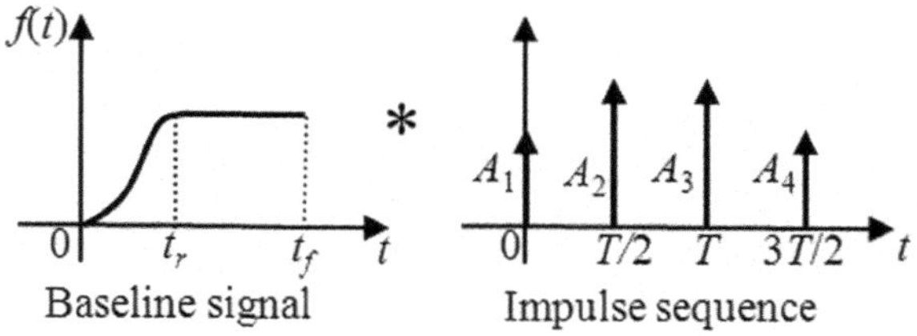

Fig. 5.22 Schematic of ZVDD input shaper

control input shown in the equation makes the residual vibration of the second-order system zero.

5.4.1.2 Variable Amplitudes ZVDD Input Shaper

Figure 5.22 shows the principle of the ZVDD input shaper, which consists of four impulses. The final signal is obtained by convolving the baseline signal convolves with the impulse sequence. Therefore, the duration of the shaped signal is increased to be $t_f + 3\,T/2$.

Amplitudes of four impulses satisfy

$$\begin{aligned}\text{ZVDD} &= \begin{bmatrix} \dfrac{A_i}{t_{A_i}} \end{bmatrix} \\ &= \begin{bmatrix} \frac{1}{1+3K+3K^2+K^3} & \frac{3K}{1+3K+3K^2+KH^3} & \frac{3K^2}{1+3K+3K^2+K^3} & \frac{K^3}{1+3K+3K^2+K^3} \\ 0 & \frac{T}{2} & T & \frac{3}{2}T \end{bmatrix} \\ &(i = 1, 2, 3, 4)\end{aligned} \tag{5.41}$$

Introduce components of $\widehat{\boldsymbol{f}}_j$ as

$$\widehat{\boldsymbol{g}}_{j,l} = \begin{bmatrix} g_{j,l} & g_{j,l+1} & g_{j,l+2} & g_{j,l+3} \end{bmatrix}^T \quad \begin{pmatrix} j = 1, 2, \ldots, m/2 \\ l = 1, 2, \ldots, k-3 \end{pmatrix} \tag{5.42}$$

Amplitudes of the scalars in $\widehat{\boldsymbol{g}}_{j,l}$ satisfy constraints of ZVDD, namely

$$\begin{aligned}\frac{g_{j,l}}{1/(1+3K+3K^2+K^3)} &= \frac{g_{j,l+1}}{3K/(1+3K+3K^2+K^3)} \\ &= \frac{g_{j,l+2}}{3K^2/(1+3K+3K^2+K^3)} \\ &= \frac{g_{j,l+3}}{K^3/(1+3K+3K^2+K^3)}\end{aligned} \tag{5.43}$$

Hence, $\widehat{g}_{j,l}$ has the expression of

$$\widehat{\boldsymbol{g}}_{j,l} = \overline{g}_{j,l}\left[\begin{array}{cccc} \frac{1}{1+3K+3K^2+K^3} & \frac{3K}{1+3K+3K^2+K^3} & \frac{3K^2}{1+3K+3K^2+K^3} & \frac{K^3}{1+3K+3K^2+K^3} \end{array}\right]^T \tag{5.44}$$

where $\overline{g}_{j,l}$ is the amplitude of the l-th component.

$\widehat{\boldsymbol{g}}_{j,l}$ can cancel residual vibration of the second-order system. The vector $\widehat{\boldsymbol{f}}_j$ is seen as a linear combination of $\widehat{\boldsymbol{g}}_{j,l}$

$$\begin{gathered}
\widehat{\boldsymbol{f}}_j(1) = f_j = \widehat{\boldsymbol{g}}_{j,1}(1) \\
\widehat{\boldsymbol{f}}_j(2) = f_{j+\frac{m}{2}} = \widehat{\boldsymbol{g}}_{j,2}(1) + \widehat{\boldsymbol{g}}_{j,1}(2) \\
\widehat{\boldsymbol{f}}_j(3) = f_{j+\frac{m}{2}} = \widehat{\boldsymbol{g}}_{j,3}(1) + \widehat{\boldsymbol{g}}_{j,2}(2) + \widehat{\boldsymbol{g}}_{j,1}(3) \\
\vdots \\
\widehat{\boldsymbol{f}}_j(i) = f_{j+(i-1)\frac{m}{2}} = \widehat{\boldsymbol{g}}_{j,i}(1) + \widehat{\boldsymbol{g}}_{j,i-1}(2) + \widehat{\boldsymbol{g}}_{j,i-2}(3) + \widehat{\boldsymbol{g}}_{j,i-3}(4) \quad (i = 4, 5, \ldots, k-3) \\
\vdots \\
\widehat{\boldsymbol{f}}_j(k-2) = f_{j+(k-3)\frac{m}{2}} = \widehat{\boldsymbol{g}}_{j,k-3}(2) + \widehat{\boldsymbol{g}}_{j,k-4}(3) + \widehat{\boldsymbol{g}}_{j,k-5}(4) \\
\widehat{\boldsymbol{f}}_j(k-1) = f_{j+(k-2)\frac{m}{2}} = \widehat{\boldsymbol{g}}_{j,k-3}(3) + \widehat{\boldsymbol{g}}_{j,k-4}(4) \\
\widehat{\boldsymbol{f}}_j(k) = f_{j+(k-1)\frac{m}{2}} = \widehat{\boldsymbol{g}}_{j,k-3}(4)
\end{gathered} \tag{5.45}$$

namely

$$\widehat{\boldsymbol{f}}_j = \begin{bmatrix} f_j \\ f_{j+m/2} \\ f_{j+m} \\ \vdots \\ f_{j+(k-2)m/2} \\ f_{j+(k-1)m/2} \end{bmatrix} = \begin{bmatrix}
1 & & & \vdots & & & \vdots & & \\
& 1 & & \vdots\,1 & & & \vdots & & \\
& & 1 & \vdots & 1 & & \vdots & & \\
& & & 1\,\vdots & & & \vdots & & \\
& & & \vdots & & \ddots & \vdots & & \\
& & & \vdots & & & \vdots\,1 & & \\
& & & \vdots & & 1 & \vdots & 1 & \\
& & & \vdots & & & 1\,\vdots & & 1 \\
& & & \vdots & & & \vdots & & 1
\end{bmatrix} \begin{bmatrix} \widehat{\boldsymbol{g}}_{j,1} \\ \widehat{\boldsymbol{g}}_{j,2} \\ \widehat{\boldsymbol{g}}_{j,3} \\ \vdots \\ \widehat{\boldsymbol{g}}_{j,k-4} \\ \widehat{\boldsymbol{g}}_{j,k-3} \end{bmatrix} \tag{5.46}$$

Inserting Eq. (5.44) into the above equation, we can obtain

$$\widehat{\boldsymbol{f}}_j = \begin{bmatrix} f_j \\ f_{j+m/2} \\ f_{j+m} \\ \vdots \\ f_{j+(k-2)m/2} \\ f_{j+(k-1)m/2} \end{bmatrix} = \begin{bmatrix} A_1 & & & & \\ A_2 & A_1 & & & \\ A_3 & A_2 & A_1 & & \\ A_4 & A_3 & A_2 & & \\ & A_4 & A_3 & & \\ & & A_4 & \ddots & A_1 \\ & & & & A_2 & A_1 \\ & & & & A_3 & A_2 \\ & & & & A_4 & A_3 \\ & & & & & A_4 \end{bmatrix} \begin{bmatrix} \overline{g}_{j,1} \\ \overline{g}_{j,2} \\ \overline{g}_{j,3} \\ \vdots \\ \overline{g}_{j,k-4} \\ \overline{g}_{j,k-3} \end{bmatrix} \tag{5.47}$$

The vector $\widehat{\boldsymbol{f}}_j$ in form of Eq. (5.33) can make residual vibration of the system to be zero. Also, it is only necessary to find a suitable one $\overline{g}_{j,l}$.

5.4.2 Numerical Simulation

To demonstrate the effectiveness of the proposed method, a three-dimensional rest-to-rest maneuver case in the Sect. 5.3.2 is simulated. The performance of the variable amplitudes ZVD and ZVDD input shapers is compared with that of the variable amplitudes ZV input shapers. Figures 5.23, 5.24 and 5.25 shows the comparison of applied controlled torques, the angular velocity trajectories and the Euler angles, respectively. After attitude maneuver, the spacecraft converges to the target attitude angles with high precision.

With the input shaper applied, control inputs on the yaw and pitch degrees-of-freedom are both optimized to be a combination of a number of pulses. Figure 5.24 shows the corresponding angular velocities of the spacecraft in y and z direction. It

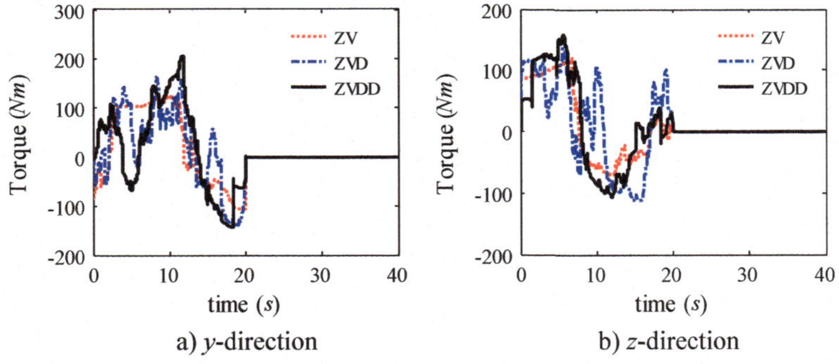

Fig. 5.23 Comparison of torques between three input shapers

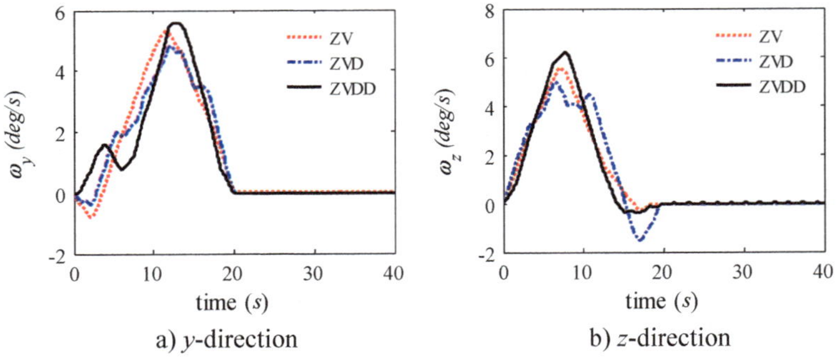

Fig. 5.24 Comparison of angular velocities between three input shapers

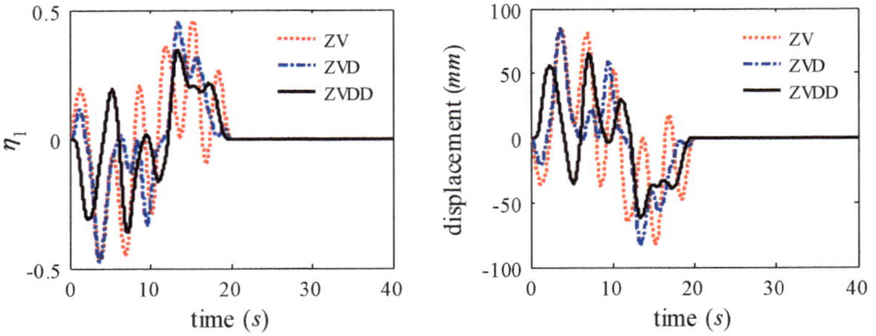

Fig. 5.25 Comparison of tip deflection between three input shapers

can be shown that all the three shaped control inputs induce very little vibration of the angular rates at the end of the control input.

Figure 5.25 illustrates the transient responses of modal coordinates and the tip displacement of the solar wing in the floating coordinate, respectively. Under the three kinds of shaped inputs, elastic vibrations are all ultimately canceled and brought to near zero. As a result, terminal oscillations of attitude angles have near zero magnitude when the controlled torque is removed.

Figures 5.26 and 5.27 illustrate oscillations of the first modal coordinate when ε equals 5% and 10% under the variable amplitudes ZVD shaper and ZVDD shaper, respectively. And Fig. 5.28 compares the tip deflection of the solar wing between the three variable amplitudes input shapers when ε equals 5%. Through comparisons of the residual vibrations, it is obvious to conclude that the robust variable amplitudes input shapers are insensitive to the frequency variation. Further, the ZVDD shaper is more robust compared to the ZV shaper.

Now the damping ratio is set to 0.01 in order to evaluate the robustness of the proposed robust maneuver strategy. Figures 5.29 and 5.30 shows time responses of the first modal coordinate when the three input shapers are used, respectively. And Fig. 5.31 compares the tip deflection of the solar wing when the error of the damping ratio is 0.01. It can be shown that the amplitude of residual vibration increases

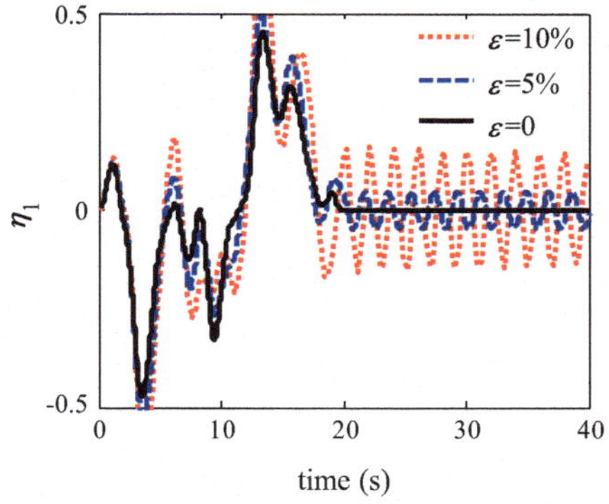

Fig. 5.26 Responses of the first modal coordinate with frequency perturbation (ZVD shaper)

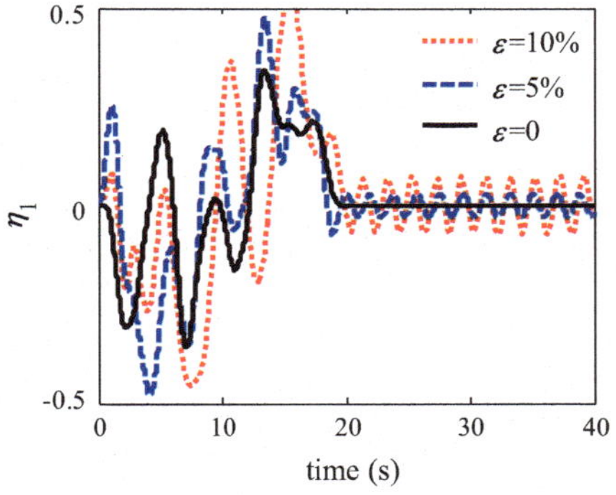

Fig. 5.27 Responses of the first modal coordinate with frequency perturbation (ZVDD shaper)

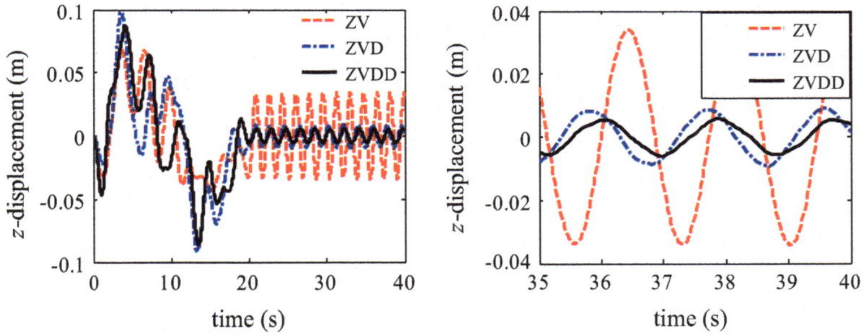

Fig. 5.28 Comparison of tip deflection between three input shapers with frequency perturbation

prominently with damping uncertainty for ZV shaper. However, the other two shapers are more robust. Hence, we can also conclude that the robust variable amplitudes input shapers are insensitive to the damping uncertainty.

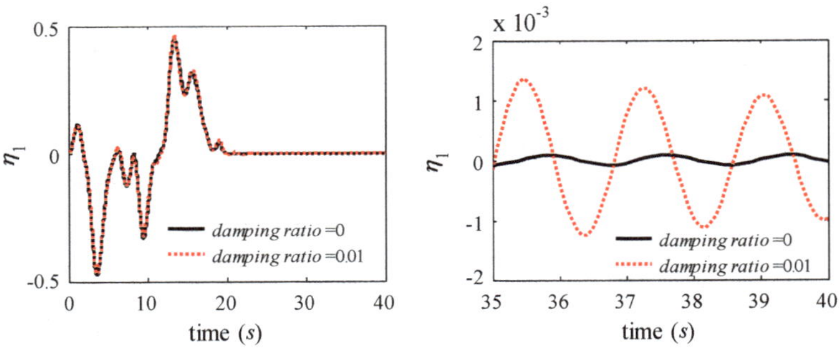

(a) complete time history (b) zoomed portion

Fig. 5.29 Responses of the first modal coordinate with damping uncertainty (ZVD shaper)

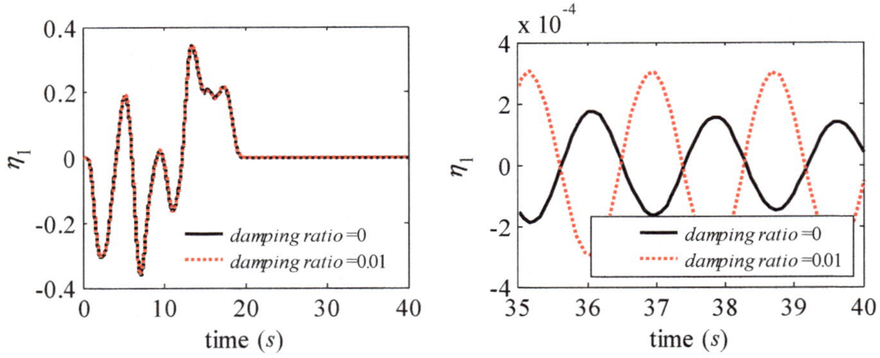

(a) complete time history (b) zoomed portion

Fig. 5.30 Responses of the first modal coordinate with damping uncertainty (ZVDD shaper)

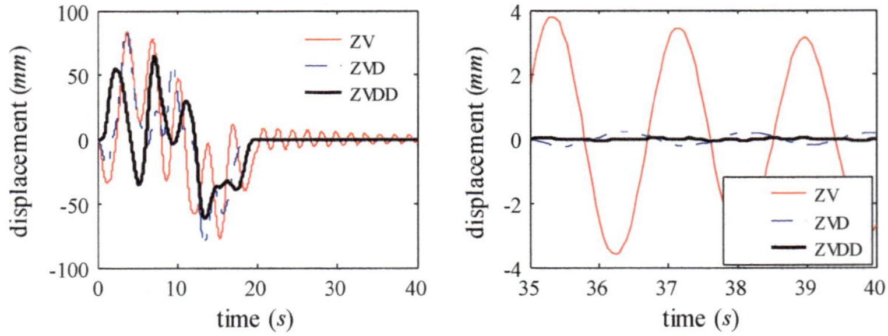

(a) complete time history (b) zoomed portion

Fig. 5.31 Comparison of tip deflection between three input shapers with damping uncertainty

5.5 Summary

Focused on the slew maneuver of flexible spacecraft with time-varying parameters and uncertainties, this chapter has designed variable amplitudes input shapers including the variable amplitudes zero vibration (ZV) input shaper, zero vibration derivative (ZVD) input shaper and the zero-vibration double-derivative (ZVDD) input shaper. And the corresponding robust attitude maneuver strategies are proposed. Simulation results show that the maneuver strategies exhibit excellent robustness and stability and solve the problem of attitude maneuvering of flexible spacecraft with variable parameters.

References

1. Farrenkopf, R. L. (2015). Optimal open-loop maneuver profiles for flexible spacecraft [J]. *Journal of Guidance & Control, 1*(6), 272–80.
2. Meirovitch, L., & Kwak, M. K. (1990). Dynamics and control of spacecraft with retargeting flexible antennas [J]. *Journal of Guidance, Control, and Dynamics, 13*(2), 241–248.
3. Loquen, T., de Plinval, H., & Cumer, C. et al. (2012). Attitude control of satellites with flexible appendages: a structured H_∞ control design. *Proceedings of the AIAA Guidance, Navigation, and Control Conference*, Minneapolis, USA, F.
4. de Souza, A. G., & de Souza, L. C. (2014). Satellite attitude control system design taking into account the fuel slosh and flexible dynamics [J]. *Mathematical Problems in Engineering.*
5. Gasbarri, P., Sabatini, M., & Pisculli, A. (2016). Dynamic modelling and stability parametric analysis of a flexible spacecraft with fuel slosh [J]. *Acta Astronautica, 127*, 141–59.
6. Khoshnood, A. M., & Kavianipour, O. (2015). Vibration suppression of fuel sloshing using subband adaptive filtering (Research Note) [J]. *International Journal of Engineering-Transactions A: Basics, 28*(10), 1507–1514.
7. Souza A. G. D., & Souza L. C. G. D. (2015). Design of satellite attitude control system considering the interaction between fuel slosh and flexible dynamics during the system parameters estimation [J]. *Applied Mechanics and Materials, 706*, 14–24.
8. Gasbarri, P., Monti, R., de Angelis, C., et al. (2014). Effects of uncertainties and flexible dynamic contributions on the control of a spacecraft full-coupled model [J]. *Acta Astronautica, 94*(1), 515–526.
9. Smith, O. J. (1957). Posicast control of damped oscillatory systems [J]. *Proceedings of the IRE, 45*(9), 1249–1255.
10. Singer, N. (1990). Seering W. *Preshaping Command Inputs to Reduce System Vibration [J]., 112*(1), 76–82.
11. Singhose, W., Derezinski, S., & Singer, N. (1996). Extra-insensitive input shapers for controlling flexible spacecraft [J]. *Journal of Guidance Control & Dynamics, 19*(2), 385–391.
12. Singhose, W. E., Seering, W. P., & Singer, N. C. (1996b). Input shaping for vibration reduction with specified insensitivity to modeling errors [J]. *Japan-USA Sym on Flexible Automation, 1*, 307–13.
13. Masoud, Z., & Alhazza, K. (2014). Frequency-modulation input shaping for multimode systems [J]. *Journal of Vibration & Control, 14*(103), 1–11.
14. Singh, T., & Heppler, G. R. (1993). Shaped input control of a system with multiple modes [J]. *Journal of Dynamic Systems Measurement & Control, 115*(3), 341–347.
15. Magee, D. P., & Book, W. J. (1992). The application of input shaping to a system with varying parameters [J]. In *Japan/USA Symposium on Flexible Automation, 1*, 519–26.

16. Magee, D. P., & Book W. J. (1992). Experimental verification of modified command shaping using a flexible manipulator [J]. *Proceedings of the 1992 First International Conference on Motion and Vibration Control*, 55355–8.
17. Pao, L. Y., & Singhose, W. E. (1995). A comparison of constant and variable amplitude command shaping techniques for vibration reduction [J]. *IEEE Conference on Control Applications*, 875–881.
18. Cho, J.-K., & Park, Y.-S. (1995). Vibration reduction in flexible systems using a time-varying impulse sequence [J]. *Robotica, 13*(03), 305–13.
19. Lee, K.-S., & Park, Y.-S. (2001). Residual vibration reduction for a flexible structure using a modified input shaping technique [J]. *Robotica, 20*(05), 553–61.
20. Otsuki, M., Shibata, S., & Yoshida, K. (2007). Robust command shaping for positioning control of time-varying flexible structure considering structured uncertainty [J]. *American Control Conference*, IEEE, pp. 4987–92.
21. Otsuki, M., Mizukami, N., & Kubota, T. (2010). Simultaneous control for position and vibration of a planetary rover with flexible structures [J]. *Advanced Robotics, 3*, 387–419.
22. Singh, T., & Vadali, S. R. (1994). Input-shaped control of three-dimensional maneuvers of flexible spacecraft [J]. *Journal of Guidance Control & Dynamics, 16*(6), 1061–1068.
23. Hu, Q., Shi, P., & Gao, H. (2007). Adaptive variable structure and commanding shaped vibration control of flexible spacecraft [J]. *Journal of Guidance Control & Dynamics, 30*(3), 804–815.
24. Orszulik, R., & Shan, J. (2011). Vibration control using input shaping and adaptive positive position feedback [J]. *Journal of Guidance Control & Dynamics, 34*(4), 1031–1044.
25. Banerjee, A. K., Pedreiro, N., & Singhose, W. E. (2001). Vibration reduction for flexible spacecraft following momentum dumping with/without slewing [J]. *Journal of Guidance, Control, and Dynamics, 24*(3), 417–427.
26. Gurleyuk, S. S. (2011). Designing unity magnitude input shaping by using PWM technique [J]. *Mechatronics, 21*(1), 125–131.
27. Mimmi, G., & Pennacchi, P. (2001). Pre-shaping motion input for a rotating flexible link [J]. *International Journal of Solids & Structures, 38*(10–13), 2009–2023.
28. Li W. P., Luo, B., & Huang, H. (2016). Active vibration control of flexible joint manipulator using input shaping and adaptive parameter auto disturbance rejection controller [J]. *Journal of Sound & Vibration, 363*, 97–125.
29. Adams, C., Potter, J., & Singhose, W. (2015). Input-shaping and model-following control of a helicopter carrying a suspended load [J]. *Journal of Guidance Control & Dynamics, 38*(1), 94–105.
30. Yang, T. S., Chen, K. S., Lee, C. C., et al. (2007). Suppression of motion-induced residual longitudinal vibration of an elastic rod by input shaping [J]. *Journal of Engineering Mathematics, 57*(4), 365–379.
31. Chen, K. S., Ou, K. S., Chen, K. S., et al. (2010). Simulations and experimental investigations on residual vibration suppression of electromagnetically actuated structures using command shaping methods [J]. *Journal of Vibration & Control, 16*(16), 1713–1734.
32. Singhose, W. E., Banerjee, A. K., & Seering, W. P. (1997). Slewing flexible spacecraft with deflection-limiting input shaping [J]. *Journal of Guidance, Control, and Dynamics, 20*(2), 291–298.
33. Sung, Y.-G. (1999). Adaptive robust vibration control with input shaping as a flexible maneuver strategy [J]. *KSME International Journal, 13*(11), 807–17.
34. Parman, S. (2013). Controlling attitude maneuvers of flexible spacecraft based on nonlinear model using combined feedback-feedforward constant-amplitude inputs [J]. In *2013 10th IEEE International Conference on Control and Automation (ICCA)*, 1584–91.
35. Setyamartana, P., & Hideo, K. (1999). Rest-to-rest attitude naneuvers and residual vibration reduction of a finite element model of flexible satellite by using input shaper [J]. *Shock and Vibration, 6*(1), 11–27.
36. Zhang, Y., & Zhang, J. (2013). Combined control of fast attitude maneuver and stabilization for large complex spacecraft [J]. *Acta Mechanica Sinica, 29*(6), 875–882.

37. Gasbarri, P., Monti, R., & Sabatini, M. (2014). Very large space structures: Non-linear control and robustness to structural uncertainties [J]. *Acta Astronautica, 93*, 252–65.
38. Miao, S., Cong, B., & Liu, X. (2013). Adaptive sliding mode control of flexible spacecraft on input shaping [J]. *Acta Aeronautica et Astronautica Sinica, 34*(8), 1906–1914.
39. Zhu, L., Ma, G., Hou, Y., et al. (2009). Adaptive sliding mode control for attitude maneuvering of flexible spacecraft [J]. *Journal of Beijing University of Technology, 35*(1), 13–18.
40. Na, S., Tang, G.-A., & Chen L.-F. (2014). Vibration reduction of flexible solar array during orbital maneuver [J]. *Aircraft Engineering and Aerospace Technology: An International Journal, 86*(2), 155–64.
41. Kim, J. J., & Agrawal, B. N. (2008). RESt-to-rest slew maneuver of three-axis rotational flexible spacecraft. *Proceedings of the 17th World Congress The International Federation of Automatic Control*, Seoul, Korea, F, 2008 [C].
42. Vaughan, J., Yano, A., & Singhose, W. (2008). Comparison of robust input shapers [J]. *Journal of Sound and Vibration, 315*, 797–815.

Chapter 6
Rigid-Flexible Coupling Control Method for Flexible Spacecraft

6.1 Introduction

For further improving the performances of attitude control methods, vibration controllers for flexible appendages have been introduced and the coupling control method has been developed. The basic idea of the coupling control method is to combine the attitude stabilization theory with a vibration controller. The attitude of the platform and elastic vibrations of flexible appendages are controlled simultaneously.

In general, frequency bands of the attitude control subsystem and the vibration control subsystem are non-overlapping or the coupling is weak. Under this circumstance, the closed loops for the two subsystems can be designed separately. And the designed controllers can comply with the requirements. As the scale of flexible appendages increases and the natural frequencies decrease, severe overlap occurs in the frequency bands of the two subsystems. One way to conceive of the controller is to decouple the two subsystems before the controller design. Quinnt and Meirovitch [1] proposed a first-order perturbation approach to separate the equations of motion into a set of equations governing rigid-body slewing of the spacecraft and a set of time-varying linear equations governing small elastic motions. Then a maneuver force distribution was developed which excites the least amount of elastic deformation of the flexible parts of the spacecraft. Kakad [2] used four Euler parameters to express the motion of a flexible spacecraft and the maneuver problem was reduced to be a system of uncoupled equations. Azadi et al. [3] divided the system dynamics into two fast and slow subsystems using singular perturbation theory.

To date the applied attitude control subsystem is mainly based on PD, LQR, sliding mode control and back-stepping control theories, etc. The flexible control subsystem is based on passive or active vibration control method and the actuators are mainly piezoelectric materials. Sales et al. [4] constructed a passive control system for flexible appendages using piezoelectric transducers. The results demonstrated that the plate vibrations levels and coupling between the flexible and rigid body

© The Author(s), under exclusive license to Springer Nature Singapore Pte Ltd. 2022
J. Wang and D.-X. Li, *Rigid-Flexible Coupling Dynamics and Control of Flexible Spacecraft with Time-Varying Parameters*,
https://doi.org/10.1007/978-981-16-5097-0_6

motions were significantly reduced during the spacecraft maneuver. Silverberg and Baruh [5] used a natural control to suppress the vibration so that the control of the elastic motion does not distort the maneuver. Then the maneuver of the platform can be designed and performed independently. Then Baruh [6] applied this method to lightweight multibody structures, and designed a vibration controller based on piezoelectric film actuators. But the decouping procedure depends on the accuracy of the system's dynamic model. And the error and uncertainties of the model were ignored. Meirovitch [7] designed both open loop and closed loop controllers, in which the former one was to cancel the environmental disturbance and the later one was utilized to control the spacecraft and flexible appendage based on the optimal control theory.

Gennaro [8] studied the active suppression of appendages during attitude tracking or large-angle maneuver. A PD controller was designed for attitude with thruster and fly wheel while the piezo-electrical materials were used to suppress the vibration of a flexible beam. Grewal and Modi [9] proposed a two-level strategy using the LQG/LTR approach and the strategy achieved good attitude control and vibration suppression behavior. Zhu et al. [10] studied a robust hybrid control design method which combines the backstepping control law with the strain rate feedback control method. Azadi et al. [11] applied an adaptive-robust control scheme to three axes maneuver of a flexible satellite, And the control method also suppressed the elastic vibration. Cui and Xu [12] designed the attitude control method via θ-D method and adopted the positive position feedback (PPF) for vibration control. Shahravi and Azimi [13] compared the performance of the collocated and non-collocated piezoceramic patches acting as sensors and actuators during attitude maneuver. Azadi et al. [3] established the dynamic model for a flexible spacecraft under three axes slewing maneuver. And they proposed an adaptive-robust control scheme for attitude control and a Lyapunov based controller for the vibration suppression of the flexible structure. Qiu [14] proposed the use of combination of positive feedback and PD control. And the experimental results proved that the proposed scheme can quickly suppress vibration.

Some literature combines the input shaping method for attitude maneuver and the active vibration control for flexible appendages. SONG and AGRAWAL [15] presented an approach which is built on the input shaping method and the pulse-width pulse-frequency (PWPF) modulator to reduce elastic vibrations during the attitude control. The closed loop sub-system of attitude feedback control employs a PD controller while the active vibration suppression sub-system uses the PPF control strategy. Experiments were conducted on a system which consists of a central hub and a L-shape flexible appendage. Based on Song's work, in order to solve the problem of model uncertainties, Hu and Ma [16, 17] proposed an approach based on PPF to vibration reduction during the maneuver by using the theory of variable structure control [18]. The attitude controller consisted of linear and noncontinuous feedback items, which ensure the sliding manifold existing and global reachable. Hu and Ma [19] programmed the maneuver trajectory based on the component synthesis vibration suppression (CSVS) method with the pulse-width pulse-frequency (PWPF) modulation. And the PPF control technique using piezoelectric materials, acted on the flexible parts. The CSVS method and the input shaping method share the similar principle to reduce the residual vibration. Then Hu [20] used the backstepping method and modal velocity feedback method for attitude control and elastic vibration. In general, depending on the idea of input shaping method, the behavior of attitude actuators is programmed, for instance, the on–off of thrusters or the output torques of wheels. Na et al. [21] proposed another way to reduce vibration through controlling the rotation at the root of the solar array using zero-placement input shaping technique.

In former research, the attitude control subsystem and the vibration control subsystem were mostly designed independently. And the stabilization of the whole system was not considered. Few literatures conducted experiments to validate the controller. In this chapter, the authors synthesize the robust H_∞ controller, the adaptive sliding mode controller, and the PPF controller to stabilize flexible spacecrafts which undergo large-angle maneuvers. The coupling methods use a robust controller or an adaptive sliding mode controller to control the attitude of the platform and use the positive position feedback controller to suppress the vibration of the flexible structure.

6.2 Robust H_∞ Attitude Control Method

When a flexible spacecraft performs a maneuver with a small angle, the nonlinear terms in its dynamic equations can be neglected. In this section, the attitude controllers are designed based on the H_∞ method for systems with constant parameters and time-varying parameters.

6.2.1 H_∞ Attitude Controller Design

6.2.1.1 Governing Equation

When the spacecraft performs a single-axis attitude maneuver or a three-axis small-angle maneuver, the formula in Eq. (2.135) is simplified to

$$\begin{aligned} \boldsymbol{J}(\gamma)\ddot{\boldsymbol{\theta}} + \boldsymbol{D}(\gamma)\ddot{\boldsymbol{\eta}} &= \boldsymbol{\tau} \\ \boldsymbol{D}(\gamma)^T\ddot{\boldsymbol{\theta}} + \ddot{\boldsymbol{\eta}} + 2\boldsymbol{\xi}\boldsymbol{\Lambda}\dot{\boldsymbol{\eta}} + \boldsymbol{\Lambda}^2\boldsymbol{\eta} &= 0 \end{aligned} \tag{6.1}$$

where $\boldsymbol{\theta}$ is the attitude angle vector and satisfies $\dot{\theta} = \boldsymbol{\omega}$. The above equation can be expressed as

$$\begin{bmatrix} J(\gamma) & D(\gamma) \\ D(\gamma)^T & I_{n\times n} \end{bmatrix}\begin{bmatrix} \ddot{\theta} \\ \ddot{q} \end{bmatrix} + \begin{bmatrix} 0 & 0 \\ 0 & 2\xi\Lambda \end{bmatrix}\begin{bmatrix} \dot{\theta} \\ \dot{q} \end{bmatrix} + \begin{bmatrix} 0 & 0 \\ 0 & \Lambda^2 \end{bmatrix}\begin{bmatrix} \theta \\ q \end{bmatrix} = \begin{bmatrix} \tau \\ 0 \end{bmatrix} \tag{6.2}$$

Define a state variable to be

$$x = \begin{bmatrix} \theta & q & \dot{\theta} & \dot{q} \end{bmatrix}^T \tag{6.3}$$

The state space representation of the spacecraft can be described as follows

$$\dot{x} = \left[\begin{array}{c|c} 0 & I_{3+n} \\ \hline -M_s^{-1}(\gamma)K_s & -M_s^{-1}(\gamma)C_s \end{array}\right]x + \begin{bmatrix} 0 \\ M_s^{-1}F \end{bmatrix}\tau \tag{6.4}$$

The matrices in the above equation can be expressed as

$$\begin{aligned} M_s(\gamma) &= \begin{bmatrix} J(\gamma) & D(\gamma) \\ D(\gamma)^T & I_n \end{bmatrix} & C_s &= \begin{bmatrix} 0 & 0 \\ 0 & 2\xi\Lambda \end{bmatrix} \\ K_s &= \begin{bmatrix} 0 & 0 \\ 0 & \Lambda^2 \end{bmatrix} & F &= \begin{bmatrix} I_3 \\ 0 \end{bmatrix} \end{aligned} \tag{6.5}$$

Assuming that the attitude angle and attitude angular velocity are measurable, the system is

$$\begin{aligned} \dot{x} &= A(\gamma)x + B(\gamma)\tau \\ y &= Cx + D\tau \end{aligned} \tag{6.6}$$

D in the above formula has a different meaning from $D(\gamma)$ in Eq. (6.1), and

$$\begin{aligned} A(\gamma) &= \left[\begin{array}{c|c} 0 & I_{3+n} \\ \hline -M_s^{-1}(\gamma)K_s & -M_s^{-1}(\gamma)C_s \end{array}\right] \\ B(\gamma) &= \begin{bmatrix} 0 \\ M_s^{-1}(\gamma)F \end{bmatrix} \quad C = \begin{bmatrix} I_3 & 0 & 0 & 0 \\ 0 & 0 & I_3 & 0 \end{bmatrix} \quad D = 0 \end{aligned} \tag{6.7}$$

6.2.1.2 H$_\infty$ Controller for System with Constant Parameters

When neglecting changes of the system parameters, an optimal feedback controller can be designed according to the H$_\infty$ design method. The standard H$_\infty$ control problem is shown in Fig. 6.1, where w is the external input signal, z is the controlled output signal, u is the control signal, and y is the measurement output signal. G is the general controlled object, and K is the controller.

The generalized control object can be expressed as

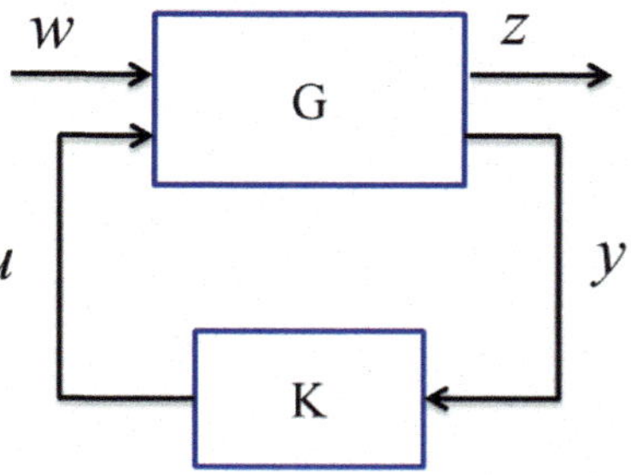

Fig. 6.1 Basic block diagram of H$_\infty$ state feedback control problem

$$G(s) = \left[\begin{array}{c|cc} A & B_1 & B_2 \\ \hline C_1 & D_{11} & D_{12} \\ C_2 & D_{21} & 0 \end{array}\right] \tag{6.8}$$

The corresponding state space realization is expressed as

$$\begin{cases} \dot{x} = Ax + B_1 w + B_2 u \\ z = C_1 x + D_{11} w + D_{12} u \\ y = C_2 x + D_{21} w + D_{22} u \end{cases} \tag{6.9}$$

Define the state feedback control law as

$$\tau = Ky \tag{6.10}$$

where K is the state feedback gain matrix, and τ is the controlled torque, which is the control signal u.

Consider the case of $D_{22} = 0$, Introduce a sufficiently small constant $\varepsilon > 0$ and construct an aided system as follows

$$G_\varepsilon(s) = \left[\begin{array}{c|cc} A & \begin{bmatrix} B_1 & \sqrt{\varepsilon} Q_b & 0 \end{bmatrix} & B_2 \\ \hline \begin{bmatrix} C_1 \\ \sqrt{\varepsilon} Q_c \\ 0 \end{bmatrix} & \begin{bmatrix} D_{11} & 0 & 0 \\ 0 & 0 & 0 \\ 0 & 0 & 0 \end{bmatrix} & \begin{bmatrix} D_{12} \\ 0 \\ \sqrt{\varepsilon} Q_{12} \end{bmatrix} \\ C_2 & \begin{bmatrix} D_{21} & 0 & \sqrt{\varepsilon} Q_{21} \end{bmatrix} & 0 \end{array}\right] \tag{6.11}$$

where

$$Q_b = I_N \quad Q_c = I_N$$

$$\boldsymbol{Q}_{12} = \boldsymbol{I}_{m_2} \quad Q_{21} = \boldsymbol{I}_{p_2} \tag{6.12}$$

And

$$\begin{aligned}
&rank\begin{bmatrix} \boldsymbol{D}_{12} \\ 0 \\ \sqrt{\varepsilon}\boldsymbol{Q}_{12} \end{bmatrix} = m_2 \\
&rank\begin{bmatrix} \boldsymbol{D}_{21} & 0 & \sqrt{\varepsilon}\boldsymbol{Q}_{21} \end{bmatrix} = p_2 \\
&rank\begin{bmatrix} \boldsymbol{A} - j\omega\boldsymbol{I} & \boldsymbol{B}_2 \\ \boldsymbol{C}_1 & \boldsymbol{D}_{12} \\ \sqrt{\varepsilon}\boldsymbol{I}_N & 0 \\ 0 & \sqrt{\varepsilon}\boldsymbol{I}_{m_2} \end{bmatrix} = N + m_2, \quad \forall\omega \in \mathbb{R} \\
&rank\begin{bmatrix} \boldsymbol{A} - j\omega\boldsymbol{I} & \boldsymbol{B}_1 & \sqrt{\varepsilon}\boldsymbol{I}_N & 0 \\ \boldsymbol{C}_2 & \boldsymbol{D}_{21} & 0 & \sqrt{\varepsilon}\boldsymbol{I}_{p_2} \end{bmatrix} = N + p_2, \quad \forall\omega \in \mathbb{R}
\end{aligned} \tag{6.13}$$

It can be seen that the system shown in Eq. (6.11) satisfies the robust controller design conditions, and the controller can be designed according to the standard H_∞ controller design process.

Consequently, the new generalized control object is

$$\boldsymbol{G}_\varepsilon(s) = \left[\begin{array}{c|cc} \boldsymbol{A} & \overline{\boldsymbol{B}}_1 & \overline{\boldsymbol{B}}_2 \\ \hline \overline{\boldsymbol{C}}_1 & \overline{\boldsymbol{D}}_{11} & \overline{\boldsymbol{D}}_{12} \\ \overline{\boldsymbol{C}}_2 & \overline{\boldsymbol{D}}_{21} & 0 \end{array}\right] \tag{6.14}$$

The H_∞ solution contains the following two Hamilton matrices

$$\begin{aligned}
\boldsymbol{H}_\infty &= \begin{bmatrix} \boldsymbol{A} & \gamma^{-2}\overline{\boldsymbol{B}}_1\overline{\boldsymbol{B}}_1^* - \overline{\boldsymbol{B}}_2\overline{\boldsymbol{B}}_2^* \\ -\overline{\boldsymbol{C}}_1^*\overline{\boldsymbol{C}}_1 & -\boldsymbol{A}^* \end{bmatrix} \\
\boldsymbol{J}_\infty &= \begin{bmatrix} \boldsymbol{A}^* & \gamma^{-2}\overline{\boldsymbol{C}}_1\overline{\boldsymbol{C}}_1^* - \overline{\boldsymbol{C}}_2\overline{\boldsymbol{C}}_2^* \\ -\overline{\boldsymbol{B}}_1\overline{\boldsymbol{B}}_1^* & -\boldsymbol{A} \end{bmatrix}
\end{aligned} \tag{6.15}$$

There exists a permissible controller

$$\boldsymbol{K} = \left[\begin{array}{c|c} \hat{\boldsymbol{A}}_\infty & -\boldsymbol{Z}_\infty\boldsymbol{L}_\infty \\ \hline \boldsymbol{F}_\infty & 0 \end{array}\right] \tag{6.16}$$

where

$$\boldsymbol{X}_\infty = Ric(\boldsymbol{H}_\infty) \qquad \boldsymbol{Y}_\infty = Ric(\boldsymbol{J}_\infty)$$

$$\begin{aligned}
\hat{A}_\infty &= A + \gamma^{-2}\overline{B}_1\overline{B}_1^*X_\infty + \overline{B}_2F_\infty + Z_\infty\overline{C}_2 \\
F_\infty &= -\overline{B}_2^*X_\infty \\
L_\infty &= -Y_\infty\overline{C}_2^* \\
Z_\infty &= \left(I - \gamma^{-2}Y_\infty X_\infty\right)^{-1}
\end{aligned} \tag{6.17}$$

6.2.1.3 H$_\infty$ Controller for System with Time-Varying Parameters

For a system with time-varying parameters, the system at a certain moment is selected as the nominal system, and the system change is regarded as an uncertainty problem. The flexible spacecraft system with variable-parameters is decomposed into a determined portion and an indeterminate portion, wherein the determined portion is a nominal system. The factor causing the change in system parameters is γ, which causes a change in the mass matrix $M_s(\gamma)$ in the system equations. Decompose the matrix into the nominal matrix M_{norm} and the uncertainty variation $M_\Delta(\gamma)$, then

$$M_s(\gamma) = M_{norm} + M_\Delta(\gamma) \tag{6.18}$$

Define a matrix

$$E(\gamma) = \begin{bmatrix} I_{3+n} & \\ & M_s(\gamma) \end{bmatrix} \tag{6.19}$$

Then

$$E(\gamma) = E_{norm} + E_\Delta(\gamma) \tag{6.20}$$

where

$$\begin{aligned}
E_{norm} &= \begin{bmatrix} I_{3+n} & \\ & M_{norm} \end{bmatrix} \\
E_\Delta(\gamma) &= \begin{bmatrix} 0_{3+n} & \\ & M_\Delta(\gamma) \end{bmatrix}
\end{aligned} \tag{6.21}$$

$E_\Delta(\gamma)$ can be expressed as a function of scalar $\delta_i(\gamma)$

$$E_\Delta(\gamma) = \sum_{i=1}^{k} \delta_i(\gamma)E_i \tag{6.22}$$

where $\delta_i(\gamma)$ needs to satisfy $|\delta_i(\gamma)|<1$, $\boldsymbol{E}_i$ is a series of matrices containing one non-zero element, then

$$\boldsymbol{E}_{norm}^{-1}\boldsymbol{E}_{\Delta}(\gamma) = \begin{bmatrix} 0_{3+n} & \\ & \boldsymbol{M}_{norm}^{-1}\boldsymbol{M}_{\Delta}(\gamma) \end{bmatrix} = \boldsymbol{L\Delta R} \tag{6.23}$$

where

$$\begin{aligned} \boldsymbol{\Delta} &= diag\left[\delta_1(\gamma)\boldsymbol{I}_1 \cdots \delta_k(\gamma)\boldsymbol{I}_k\right] \\ \boldsymbol{L} &= [\boldsymbol{L}_1 \cdots \boldsymbol{L}_k] \\ \boldsymbol{R} &= \left[\boldsymbol{R}_1^T \cdots \boldsymbol{R}_k^T\right] \end{aligned} \tag{6.24}$$

Defining the structural uncertainty as

$$\boldsymbol{w} = \boldsymbol{\Delta} z = -\boldsymbol{\Delta R}\dot{\boldsymbol{x}} \tag{6.25}$$

Therefore, a system with uncertainty is expressed as

$$\begin{aligned} \dot{\boldsymbol{x}} &= \boldsymbol{A}_{norm}\boldsymbol{x} + \boldsymbol{B}_{norm}\boldsymbol{\tau} + \boldsymbol{L}\boldsymbol{w} \\ z &= -\boldsymbol{R}\boldsymbol{A}_{norm}\boldsymbol{x} - \boldsymbol{R}\boldsymbol{B}_{norm}\boldsymbol{\tau} - \boldsymbol{R}\boldsymbol{L}\boldsymbol{w} \\ \boldsymbol{y} &= \boldsymbol{C}\boldsymbol{x} + \boldsymbol{D}\boldsymbol{\tau} \end{aligned} \tag{6.26}$$

where

$$\boldsymbol{A}_{norm} = \left[\begin{array}{c|c} 0 & \boldsymbol{I}_{3+n} \\ \hline -\boldsymbol{M}_{norm}^{-1}\boldsymbol{K}_s & -\boldsymbol{M}_{norm}^{-1}\boldsymbol{C}_s \end{array}\right] \tag{6.27}$$

Defining $\boldsymbol{e}$ as the attitude angle deviation, the form of the system $\boldsymbol{G}$ to be controlled is

$$\begin{aligned} \dot{\boldsymbol{x}} &= \boldsymbol{A}_{norm}\boldsymbol{x} + \left[0\ \boldsymbol{B}_{norm}\ \boldsymbol{L}\right]\begin{bmatrix} \boldsymbol{d} \\ \boldsymbol{\tau} \\ \boldsymbol{w} \end{bmatrix} \\ \begin{bmatrix} \dot{\boldsymbol{e}} \\ \boldsymbol{e} \\ z \end{bmatrix} &= \begin{bmatrix} -\left[0\ 0\ \boldsymbol{I}_3\ 0\right] \\ -\left[\boldsymbol{I}_3\ 0\ 0\ 0\right] \\ -\boldsymbol{R}\boldsymbol{A}_{norm} \end{bmatrix}\boldsymbol{x} + \begin{bmatrix} 0 & 0 & 0 \\ \boldsymbol{I}_3 & 0 & 0 \\ 0 & -\boldsymbol{R}\boldsymbol{B}_{norm} & -\boldsymbol{R}\boldsymbol{L} \end{bmatrix}\begin{bmatrix} \boldsymbol{d} \\ \boldsymbol{\tau} \\ \boldsymbol{w} \end{bmatrix} \end{aligned} \tag{6.28}$$

In the equation, $\boldsymbol{d}$ is the attitude angle reference input. The system turns into a standard H_∞ problem.

6.2.2 *Numerical Simulation with the H_∞ Attitude Controller*

6.2.2.1 Parameters of the H_∞ Controller

Figure 6.2 shows the adopted model for simulation, which consists of a main hub and a flexible plate bonded with PZT layers. The flexible plate is fixed to the hub through a yoke. The moment of inertia of the system is selected as 10 kg·m^2. The body coordinate of the hub is denoted as *oxyz* and the body coordinate for the plate is defined as $o_s x_s y_s z_s$. The coordinate axes of the three directions are respectively parallel to the three coordinate axes of the coordinate system *oxyz*, and the position vector of the origin o_s in the coordinate system *oxyz* is [0.4, 0, 0]m. The flexible plate material is aluminum, the size is 1500 mm × 400 mm × 1.4 mm, the mass is 2.27 kg, and the moment of inertia about the *y*-axis is 3.97 kg·m^2.

The hub can rotate about the *y* axis so that only the flexural modes in y_s direction of the plate are coupled with the hub's attitude. Therefore, the first four flexural modes in y_s direction for the plate and the hub-plate system are obtained, as given in Table 6.1. Compared to the natural frequencies of the plate, the corresponding natural frequencies of the hub-plate system are higher, especially for lower modes. The table also gives the coupling coefficient between the flexible modes and the attitude. It can be noted that with the decrease of the modal index, the coupling coefficient between the flexible mode and the attitude becomes larger, which indicates that the influence on the angular velocity becomes greater.

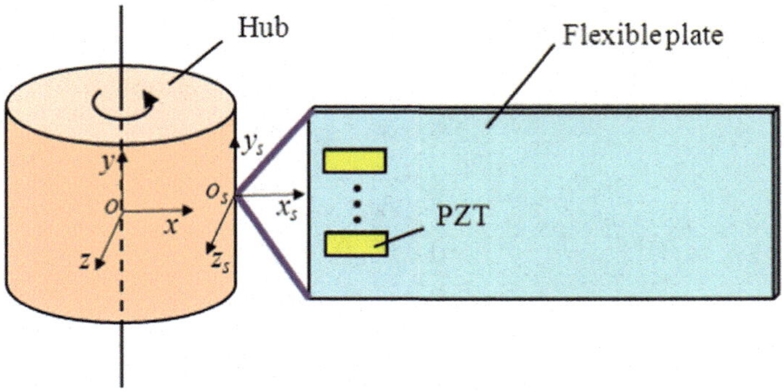

Fig. 6.2 Schematic diagram of the Hub-Plate system

Table 6.1 Natural frequencies (Hz) and coupling ratio of the Plate and Hub-plate

Modal index	Plate	Hub-plate	Coupling coefficient
1	0.520	0.645	1.87
2	3.26	3.33	−0.537
3	9.17	9.23	−0.268
4	18.1	18.1	−0.176

The central hub performs an attitude maneuver of 10 deg. At the initial time the attitude angle is 10 deg and the attitude angular velocity is zero. At the end time the attitude angle needs to be zero. The range of controlled torque that the actuator can provide is $[-0.1, 0.1]$Nm. The first three modes of the flexible plate are retained, and the modal damping ratio is assumed to be zero. The PPF controller parameters are selected as $\boldsymbol{\Lambda}_c = \boldsymbol{\Lambda}, \quad \boldsymbol{\xi}_c = 0.4$.

The simulation assumes that the natural frequencies and the coupling coefficient between the flexible plate and the central rigid body is constant. Due to the rotation of the flexible plate, the moment of inertia of the system is time-varying, and the variation law is $J = (10 + \sin\gamma)\text{kg·m}^2$. Therefore, the system moment of inertia varies within the range of [9, 11] kg·m^2, the nominal value is 10 kg·m^2, and the uncertainty is $\pm 10\%$.

The nominal matrix and uncertainty matrix of the system mass matrix are

$$\boldsymbol{M}_{norm} = \begin{bmatrix} 10 & 1.8725 & -0.5367 & -0.2677 \\ 1.8725 & 1 & 0 & 0 \\ -0.5367 & 0 & 1 & 0 \\ -0.2677 & 0 & 0 & 1 \end{bmatrix}$$

$$\boldsymbol{M}_{\Delta}(\gamma) = \begin{bmatrix} \sin\gamma & 0 & 0 & 0 \\ 0 & 0 & 0 & 0 \\ 0 & 0 & 0 & 0 \\ 0 & 0 & 0 & 0 \end{bmatrix} \tag{6.26}$$

Therefore,

$$\boldsymbol{E}_{norm}^{-1}\boldsymbol{E}_{\Delta}(\gamma) = \begin{bmatrix} 0\,0\,0\,0 & 0 & 0\,0\,0 \\ 0\,0\,0\,0 & 0 & 0\,0\,0 \\ 0\,0\,0\,0 & 0 & 0\,0\,0 \\ 0\,0\,0\,0 & 0 & 0\,0\,0 \\ 0\,0\,0\,0 & 0.163\sin\gamma & 0\,0\,0 \\ 0\,0\,0\,0 & -0.3053\sin\gamma & 0\,0\,0 \\ 0\,0\,0\,0 & 0.0875\sin\gamma & 0\,0\,0 \\ 0\,0\,0\,0 & 0.0436\sin\gamma & 0\,0\,0 \end{bmatrix} \tag{6.30}$$

Take $\delta_1(\gamma) = 0.163$, $\delta_2(\gamma) = -0.3053$, $\delta_3(\gamma) = 0.0875$, $\delta_4(\gamma) = 0.0436$, meet $|\delta_i(\gamma)| < 1$, and

$$
\boldsymbol{\Delta} = \begin{bmatrix} 0.163 & & & \\ & -0.3053 & & \\ & & 0.0875 & \\ & & & 0.0436 \end{bmatrix}
$$

$$
\boldsymbol{L} = \begin{bmatrix} \boldsymbol{0}_{4\times 4} \\ \boldsymbol{I}_{4\times 4} \end{bmatrix} \tag{6.31}
$$

$$
\boldsymbol{R} = \left[\begin{array}{c|cccc} & 1 & 0 & 0 & 0 \\ \boldsymbol{0}_{4\times 4} & 1 & 0 & 0 & 0 \\ & 1 & 0 & 0 & 0 \\ & 1 & 0 & 0 & 0 \end{array}\right]
$$

Δ, ***L*** and ***R*** in the above formula satisfy the Eq. (6.23). The calculated feedback controller ***K*** is

$$
\boldsymbol{K} = \left[\begin{array}{cccccccc|cc}
-1.40 & 0 & 0.002 & -0.024 & 0.181 & 0 & 0 & 0 & 17.9 & 10.5 \\
0.012 & 0.008 & -0.053 & 0.781 & -1.45 & 1.00 & -0.003 & -0.005 & -0.166 & 18.5 \\
-0.006 & -0.007 & 0.043 & -0.644 & 0.312 & -0.002 & 1.00 & 0.004 & 0.087 & -3.95 \\
-0.002 & -0.019 & 0.129 & -1.92 & 0.210 & -0.006 & 0.007 & 1.01 & 0.043 & -2.57 \\
-0.900 & 1.01 & -6.76 & 100 & -11.3 & 0.337 & -0.382 & -0.655 & 10.5 & 139 \\
0.421 & -12.5 & 12.4 & -184 & 12.1 & -0.686 & 0.703 & 1.20 & -3.49 & -143 \\
-0.104 & 0.550 & -423 & 54.5 & -19.1 & 0.183 & -0.616 & -0.355 & 0.768 & 241 \\
-0.049 & 0.290 & -1.94 & -3290 & -63.9 & 0.096 & -0.109 & -1.34 & 0.328 & 815 \\
\hline
-0.034 & -1.09 & 14.4 & 117 & -0.214 & 0.149 & -0.162 & -0.283 & 0 & 0
\end{array}\right] \tag{6.32}
$$

(6.32)

6.2.2.2 Simulation Results

Using the robust controller, the input controlled torque and the dynamic responses of the system are shown in Figs. 6.3, 6.4, 6.5, 6.6 and 6.7. At 25.6 s, the attitude angle reached the target and the deviation is reduced to 0.01 deg. The amplitude of the elastic vibration excited by maneuver is 18.4 mm. At 15 s after the start of the control, the vibration amplitude is attenuated to 0.82 mm due to the damping and compensation of the controller.

6.3 Adaptive Sliding Mode Attitude Control Method

In this section, an adaptive sliding mode control method is proposed for the attitude control of a spacecraft with time-varying parameters. The adaptive sliding mode surface is defined by the modified Rodriguez parameter (MRP), and the model is

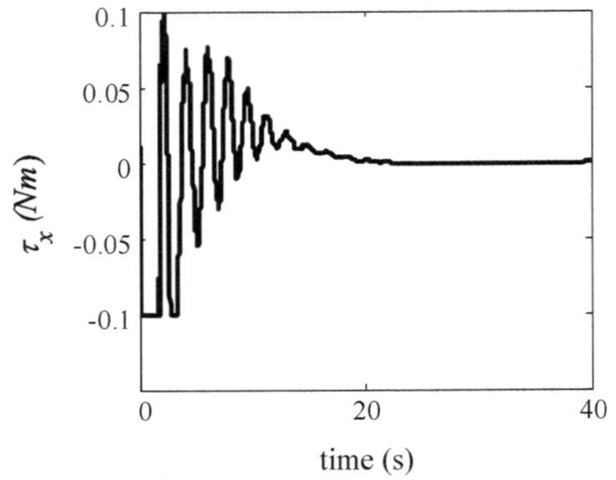

Fig. 6.3 Controlled torque input curve

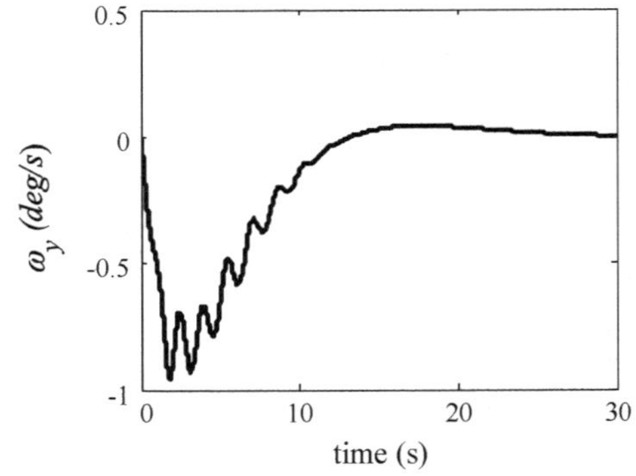

Fig. 6.4 Angular velocity response curve of central rigid body attitude

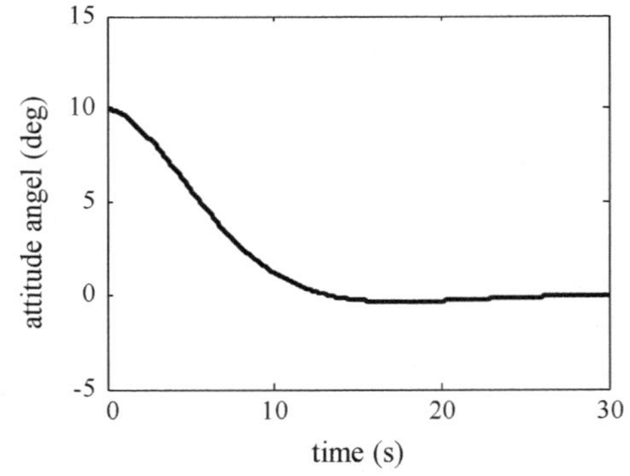

Fig. 6.5 Central rigid body attitude angle response curve

designed considering the uncertainty. Numerical simulations verify the effectiveness of the control method.

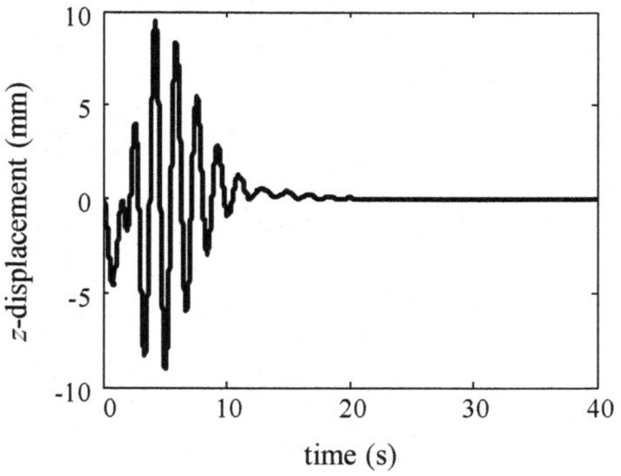

Fig. 6.6 Flexural vibration response curve at the top of flexible plate

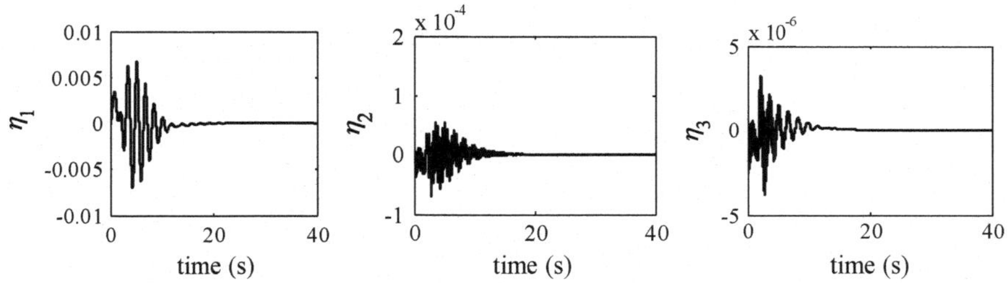

Fig. 6.7 Modal coordinate response of flexible plate

6.3.1 Adaptive Sliding Mode Controller

Use MRP to describe the attitude motion of the spacecraft, and the dynamic equation of the flexible spacecraft is

$$\begin{aligned} &\dot{\boldsymbol{\sigma}} = \boldsymbol{G}(\boldsymbol{\sigma})\boldsymbol{\omega}, \quad \boldsymbol{\sigma}(0) = \boldsymbol{\sigma}_0 \\ &\boldsymbol{J}(t)\dot{\boldsymbol{\omega}} + \boldsymbol{D}(t)\ddot{\eta} + \boldsymbol{\omega} \times (\boldsymbol{J}\boldsymbol{\omega} + \boldsymbol{D}(t)\dot{\eta}) = \boldsymbol{\tau} \\ &\boldsymbol{D}(t)^T\dot{\boldsymbol{\omega}} + \ddot{\eta} + 2\boldsymbol{\xi}\boldsymbol{\Lambda}\dot{\eta} + \boldsymbol{\Lambda}^2\eta = 0 \end{aligned} \tag{6.33}$$

6.3.1.1 Sliding Manifold

The errors of MRP and angular velocities are defined as

$$\begin{aligned} &\boldsymbol{\sigma}_e(t) = \boldsymbol{\sigma}(t) - \boldsymbol{\sigma}_d(t) \\ &\boldsymbol{\omega}_e(t) = \boldsymbol{\omega}(t) - \boldsymbol{\omega}_d(t) \end{aligned} \tag{6.34}$$

where $\boldsymbol{\sigma}_d$ is the desired reference trajectory, and $\boldsymbol{\omega}_d$ is the desired reference angular velocity.

The state trajectories move onto a sliding manifold ($\boldsymbol{s} = 0$), where $\boldsymbol{s}$ is given by

$$\boldsymbol{s}(t) = \left[\boldsymbol{\omega}(t) - \hat{\boldsymbol{\omega}}(t)\right] + \Lambda \boldsymbol{G}^{-1}(\boldsymbol{\sigma})\boldsymbol{\sigma}_e(t) \tag{6.35}$$

where $\boldsymbol{\lambda}$ is a diagonal matrix with positive elements, and the $\hat{\boldsymbol{\omega}}(t)$ is defined related to $\boldsymbol{\sigma}_d$ as

$$\hat{\boldsymbol{\omega}}(t) = \boldsymbol{G}^{-1}(\boldsymbol{\sigma})\dot{\boldsymbol{\sigma}}_d(t) \tag{6.36}$$

Theorem 1 After the states of the system arrive the sliding manifold in Eq. (6.35), the attitude tracking error $\boldsymbol{\sigma}_e \rightarrow 0$ and $\boldsymbol{\omega}_e \rightarrow 0$ as $t \rightarrow \infty$.

Proof When the states of the system arrive the sliding manifold, $\boldsymbol{s}$ satisfies

$$\boldsymbol{s}(t) = \left[\boldsymbol{\omega}(t) - \hat{\boldsymbol{\omega}}(t)\right] + \Lambda \boldsymbol{G}^{-1}(\boldsymbol{\sigma})\boldsymbol{\sigma}_e(t) = 0 \tag{6.37}$$

Multiplying by $\boldsymbol{G}(\boldsymbol{\sigma})$ and inserting Eqs. (6.33) and (6.36) result in

$$[\dot{\boldsymbol{\sigma}}(t) - \dot{\boldsymbol{\sigma}}_d(t)] + \Lambda \boldsymbol{\sigma}_e(t) = 0 \tag{6.38}$$

or

$$\dot{\boldsymbol{\sigma}}_e + \Lambda \boldsymbol{\sigma}_e(t) = 0 \tag{6.39}$$

Choose the Lyapunov function candidate as

$$V_{\sigma_e} = \frac{1}{2}\boldsymbol{\sigma}_e^T \boldsymbol{\sigma}_e \tag{6.40}$$

Differentiating Eq. (6.40) and it follows that $\dot{V}_{\sigma_e}$ satisfies

$$\dot{V}_{\sigma_e} = \boldsymbol{\sigma}_e^T \dot{\boldsymbol{\sigma}}_e = -\Lambda \boldsymbol{\sigma}_e^T \boldsymbol{\sigma}_e \leq 0 \tag{6.41}$$

If $V_{\sigma_e} \neq 0$ then $\dot{V}_{\sigma_e} < 0$, and the attitude tracking error $\boldsymbol{\sigma}_e$will decrease. If $\dot{V}_{\sigma_e} = 0$ then $\boldsymbol{\sigma}_e = 0$ and from Eqs. (6.37) and (6.36) also $\left[\boldsymbol{\omega}(t) - \hat{\boldsymbol{\omega}}(t)\right] = 0$, and $\left[\hat{\boldsymbol{\omega}}(t) - \boldsymbol{\omega}_d(t)\right] = 0$. So the angular velocities error $\boldsymbol{\omega}_e$=0.

From Eq. (6.39), the MRP error satisfies

$$\boldsymbol{\sigma}_e(t) = \boldsymbol{\sigma}_e(t_h)e^{-\Lambda(t-t_h)} \tag{6.42}$$

We can conclude that after the states of the system arrive the sliding manifold, the convergence rate of the MRP error $\boldsymbol{\sigma}_e$ depends on the coefficient $\boldsymbol{\lambda}$.

6.3.1.2 Control Law

After the design of the sliding surface, the control law is established to ensure the sliding manifold to be reachable.

Construct a Lyapunov function for the closed-loop system

$$V_s = \frac{1}{2} \boldsymbol{s}^T \boldsymbol{J} \boldsymbol{s} \tag{6.43}$$

The inertial matrix $\boldsymbol{J}$ is symmetric and positive. Hence, V_s satisfies

$$V_s \geq 0 \tag{6.44}$$

The derivative of the Lyapunov function is

$$\begin{aligned}
\dot{V}_s &= \boldsymbol{s}^T \boldsymbol{J} \dot{\boldsymbol{s}} \\
&= \boldsymbol{s}^T \boldsymbol{J} \left[\dot{\boldsymbol{\omega}}(t) - \dot{\hat{\boldsymbol{\omega}}}(t) + \Lambda \dot{\boldsymbol{G}}^{-1}(\sigma)\sigma_e(t) + \Lambda \boldsymbol{G}^{-1}(\sigma)\dot{\boldsymbol{\sigma}}_e(t) \right] \\
&= s^T [\boldsymbol{\tau} - \boldsymbol{\omega} \times \boldsymbol{J}\boldsymbol{\omega} - \boldsymbol{J}\dot{\hat{\boldsymbol{\omega}}}(t) + \Lambda \boldsymbol{J} \dot{\boldsymbol{G}}^{-1}(\sigma)\sigma_e(t) + \Lambda \boldsymbol{J} \boldsymbol{G}^{-1}(\sigma)\dot{\boldsymbol{\sigma}}_e(t) + \ldots \\
&\ldots - \boldsymbol{\omega} \times \boldsymbol{D}(t)\dot{\boldsymbol{\eta}} - \boldsymbol{D}(t)\ddot{\boldsymbol{\eta}}]
\end{aligned} \tag{6.45}$$

The torques on the platform of the spacecraft induced by the flexible appendages are taken as disturbance

$$\boldsymbol{\tau}_d = \left[\tau_{d,1}, \tau_{d,2}, \tau_{d,3}\right]^T = -\boldsymbol{\omega} \times \boldsymbol{D}(t)\dot{\boldsymbol{\eta}} - \boldsymbol{D}(t)\ddot{\boldsymbol{\eta}} \tag{6.46}$$

In a physical spacecraft, the platform's rotation would excite vibrations of lower modes for flexible appendages. And amplitudes for the multi-mode vibrations are generally limited. Moreover, the excited elastic oscillations of the flexible appendages would decrease with respect to time with structural damping or active control forces. Hence, both the deformation and velocity of the flexible appendages are assumed to be limited. So the disturbance induced by the appendages is limited and there exists

$$\left|\tau_{d,i}\right| \leq \hat{\tau}_{d,i} \qquad (i = 1, 2, 3) \tag{6.47}$$

where $|\cdot|$ denotes the absolute value of a scalar.

Define the control law as

$$\boldsymbol{\tau} = \boldsymbol{\omega} \times \boldsymbol{J}\boldsymbol{\omega} + \boldsymbol{J}\dot{\hat{\boldsymbol{\omega}}}(t) - \Lambda \boldsymbol{J} \dot{\boldsymbol{G}}^{-1}(\sigma)\sigma_e(t) - \Lambda \boldsymbol{J} \boldsymbol{G}^{-1}(\sigma)\dot{\boldsymbol{\sigma}}_e(t) + \boldsymbol{\tau}' \tag{6.48}$$

where

$$\boldsymbol{\tau}' = \left[\tau_1' \ \tau_2' \ \tau_3'\right]^T$$

$$\tau_i' = -k_i \cdot \mathrm{sgn}(s_i), \qquad (i = 1, 2, 3) \tag{6.49}$$

Sgn(·) denotes the sign function. Inserting Eqs. (6.48) into (6.45) induces

$$\begin{aligned} \dot{V}_s &= \boldsymbol{s}^T\left(\boldsymbol{\tau}' + \boldsymbol{\tau}_d\right) \\ &= -\sum_{i=1}^{3} k_i |s_i| \left[1 - \frac{\tau_{d,i}}{k_i} \cdot \frac{s_i}{|s_i|}\right] \end{aligned} \tag{6.50}$$

k_i is selected to satisfy the inequalities $k_i \geq \hat{\tau}_{d,i}$ (i = 1,2,3). Hence, $\dot{V}_s \leq 0$. If $s \neq 0$, $V_s > 0$ and $\dot{V}_s < 0$. Then the state of the system is moved to the sliding manifold. Therefore, the control law globally asymptotically stabilizes the error-based spacecraft system at the origin.

The control law in Eq. (6.48) is noncontinuous and would induce chattering because of the existence of the function sgn(·). In order to eliminate chattering, the saturation function $sat(s_i, \epsilon)$ is introduced to instead the sign function

$$sat(s_i,) = \begin{cases} \mathrm{sgn}(s_i), & |s_i| > \\ \dfrac{s_i}{}, & |s_i| \leq \end{cases} \tag{6.51}$$

After the saturation function is included, the state of the system would move in the sliding surface boundary layer $|s_i| \leq \epsilon$ instead of to be strictly restricted on the sliding manifold $s = 0$. Hence, the controlled torque is continuous and the chattering is lowered.

6.3.1.3 Control Law with Parametric Uncertainties

When the system parameters change or there are uncertainties, for example, the rotation of the flexible structure or the consumption of fuel causes the system moment of inertia to change, the robustness of the control system needs to be considered. Assuming that the nominal moment of inertia of the system is $\boldsymbol{J}_{nom}$, the difference between the actual and the normalized value is defined as

$$\Delta \boldsymbol{J} = \boldsymbol{J} - \boldsymbol{J}_{nom} \tag{6.52}$$

Then the control law is defined as

$$\boldsymbol{\tau} = \boldsymbol{\omega} \times \boldsymbol{J}_{nom}\boldsymbol{\omega} + \boldsymbol{J}_{nom}\dot{\hat{\boldsymbol{\omega}}}(t) - \Lambda \boldsymbol{J}_{nom}\dot{\boldsymbol{G}}^{-1}(\sigma)\sigma_e(t) - \Lambda \boldsymbol{J}_{nom}\boldsymbol{G}^{-1}(\sigma)\dot{\boldsymbol{\sigma}}_e(t) + \boldsymbol{\tau}' \tag{6.53}$$

Equation (6.45) becomes

$$\begin{aligned}\dot{V}_s =& \boldsymbol{s}^T\left(\boldsymbol{\tau}' + \boldsymbol{\tau}_d + \boldsymbol{\delta}\right) \\ =& -\sum_{i=1}^{3} k_i|s_i|\left[1 - \frac{\tau_{d,i} + \delta_i}{k_i} \cdot \frac{s_i}{|s_i|}\right]\end{aligned} \tag{6.54}$$

where

$$\boldsymbol{\delta} = \boldsymbol{\omega} \times \Delta \boldsymbol{J}\boldsymbol{\omega} + \Delta \boldsymbol{J}\dot{\hat{\boldsymbol{\omega}}}(t) - \Lambda\Delta\boldsymbol{J}\dot{\boldsymbol{G}}^{-1}(\boldsymbol{\sigma})\boldsymbol{\sigma}_e(t) - \Lambda\Delta\boldsymbol{J}\boldsymbol{G}^{-1}(\boldsymbol{\sigma})\dot{\boldsymbol{\sigma}}_e(t) \tag{6.55}$$

Also, the uncertainty of the spacecraft moment of inertia is usually limited. Therefore, in the above equation, independent variables of $\boldsymbol{\delta}$ are all limited. So there exists $\hat{\delta}_i (i = 1, 2, 3\ )$ such that the following inequality satisfies

$$|\delta_i| \leq \hat{\delta}_i \qquad (i = 1, 2, 3) \tag{6.56}$$

In order to ensure states of the system move to the sliding manifold, k_i is selected to satisfy $k_i \geq \hat{\tau}_{d,i} + \hat{\delta}_i$ $(i = 1, 2, 3)$.

6.3.2 *Simulation with Adaptive Sliding Mode Control Method*

In order to verify the effectiveness of the adaptive sliding mode control method, this section uses two models of the hub-plate system and the flexible spacecraft system for numerical simulation. The hub-plate system is used for single-axis attitude maneuver, and the flexible spacecraft is used for three-axis attitude maneuver.

6.3.2.1 Hub-Plate System

The model uses parameters in Sect. 6.2.2.1. The hub rotates around the *y*-axis. The first four out-of-plane bending modes of the flexible plate are considered, and the modal damping ratio is set to 0.01. Two cases are analyzed. At the initial moment the attitude angle is 60 deg, the corresponding MRP parameter is 0.2680. The target attitude angle is 0, and the corresponding MRP is 0. Parameters of adaptive sliding mode controller are $\lambda = 0.2$, $k = 0.1$, $\epsilon = 0.002$.

Figures 6.8, 6.9, 6.10, 6.11, 6.12 and 6.13 show the control input calculated by the adaptive sliding mode controller and the responses of the system when the system performs a maneuver of 60°. About 19.8 s after the start of the maneuver, the system state enters the boundary layer of the sliding mode surface, and correspondingly the controlled torque changes greatly at this time. About 40 s after the start of the maneuver, the attitude angular velocity tends to zero, and the central hub reaches the target attitude. From Fig. 6.13, the central hub excites the elastic vibration of the flexible plate. The vibration amplitude in the flexural direction (the normal direction

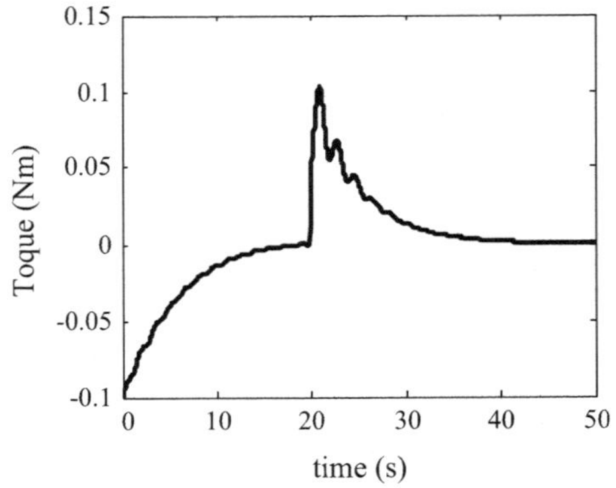

Fig. 6.8 Controlled torque input curve

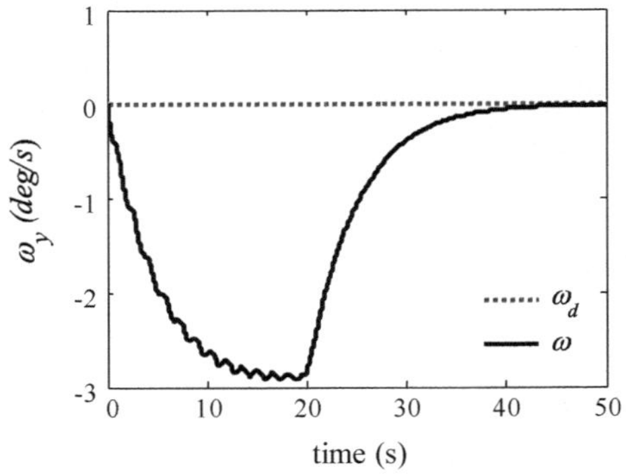

Fig. 6.9 Angular velocity response curve of central rigid body attitude

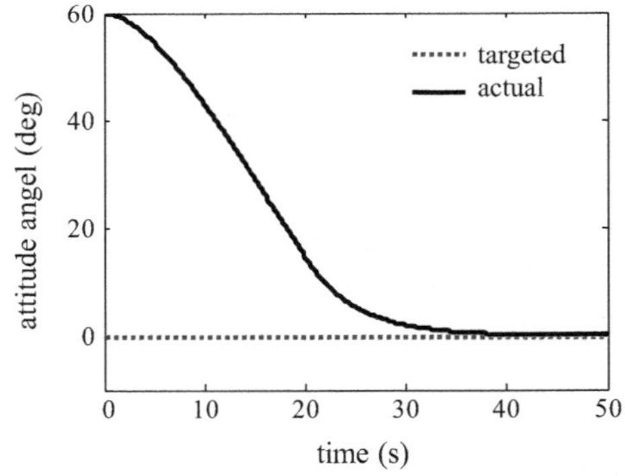

Fig. 6.10 Central rigid body attitude angle response curve

of the flexible plate plane) is 4.2 mm at the beginning, and then due to the damping of the flexible plate and the compensation of the controlled torque, the elastic vibration of the flexible plate is gradually attenuated.

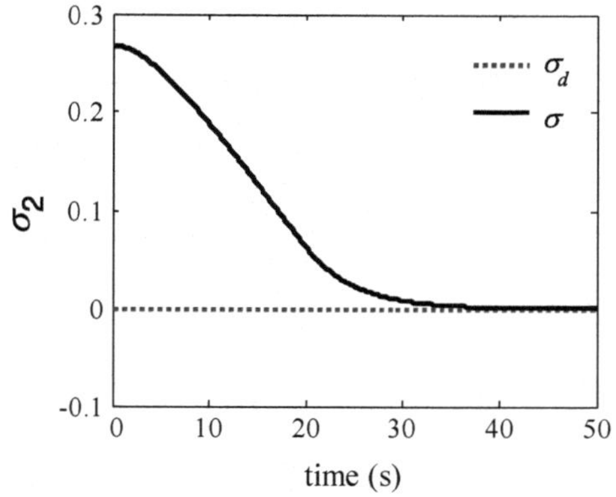

Fig. 6.11 MRP response curve of center rigid body

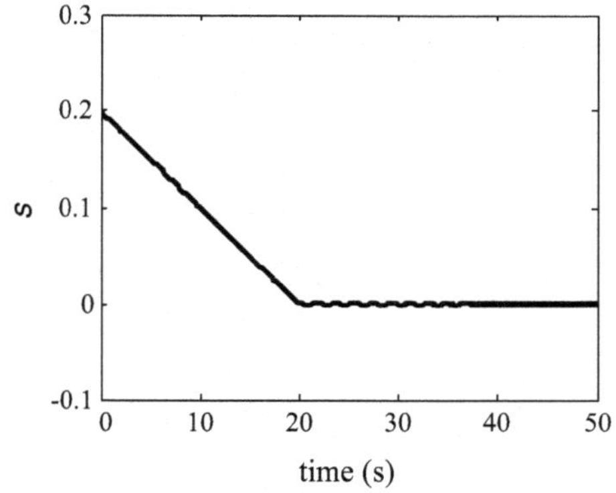

Fig. 6.12 Adaptive sliding mode controller switching function curve

Fig. 6.13 Radial vibration response curve at the top of flexible plate

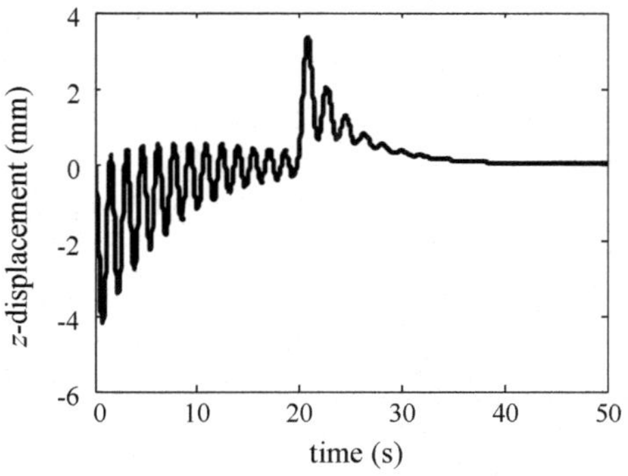

Table 6.2 The initial and target values of spacecraft attitude

Parameter	Initial value	Target value
Euler angle	[0;0;0]	[60; 45; 30]deg
Quaternion	[0;0;0;1]	[0.361; 0.440; 0.392; 0.724]
MRP	[0;0;0]	[0.209; 0.255; 0.227]
Attitude angular velocity	[0.05; −0.2; 0.05]deg/s	[0; 0; 0]

6.3.2.2 Flexible Spacecraft System

(1) **Simulation model**

The model uses the parameters in Sect. 6.3.1. The spacecraft platform performs a three-axis attitude maneuver. The Euler angle of the platform is [0;0;0] at the initial moment, and the target value is [60;45;30]deg. The corresponding quaternion and MRP are shown in Table 6.2. Considering the first six modes of the solar wing, the modal damping ratio is set to 0. Parameters of adaptive sliding mode controller are $\lambda = 0.05$, $k = 50$, $\epsilon = 0.002$.

The speed of the solar wing is non-zero, and it rotates around the xs axis at an angular velocity of $\Omega = 2$ deg/s, and the initial time ($t = 0$) $\gamma = 0°$.

The nominal values of the moment of inertia of the system are respectiely taken as

$$\boldsymbol{J}_{norm} = \begin{bmatrix} 2774 & & \\ & 7241.4 & -53.7 \\ & -53.7 & 5641.4 \end{bmatrix} \text{kg} \cdot \text{m}^2 \tag{6.57}$$

The nominal value of the moment of inertia of the system is the value when $\gamma = 45°$.

Figures 6.14, 6.15, 6.16, 6.17, 6.18, 6.19 and 6.20 show the system dynamic response of the attitude maneuver using the adaptive sliding mode control method when the solar wing rotates. About 100 s after the start of the maneuver, the attitude angular velocity tends to zero, and the parameters (MRP, Euler angle) representing the attitude of the spacecraft have reached the target value. The excited elastic vibration of the solar wing decays to zero with time. The results show that the control method exhibits strong adaptability under the condition of time-varying parameters.

6.4 Coupling Control Method for Flexible Spacecraft

In the last two sections, the robust H_∞ and adaptive sliding mode controllers are designed to control the attitude of a flexible spacecraft. The simulation results show that serious elastic vibration of the flexible structure is stimulated during the attitude

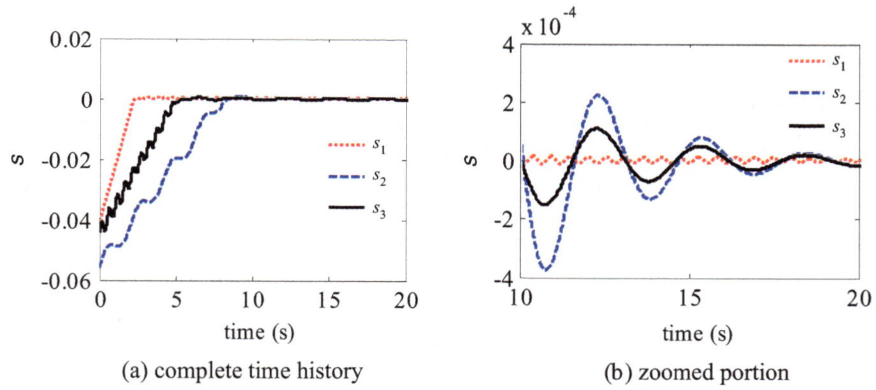

Fig. 6.14 Sliding manifold

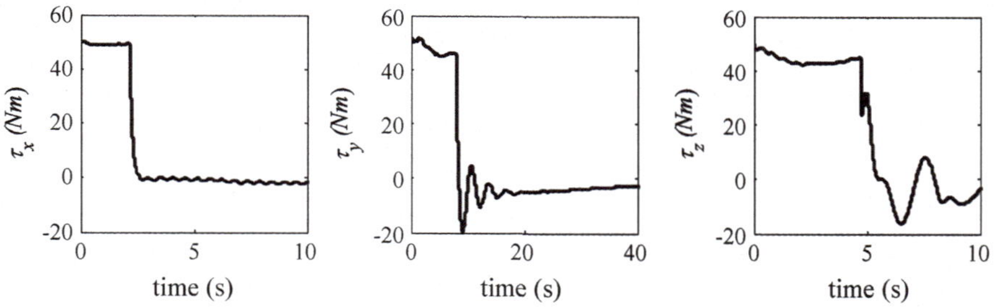

Fig. 6.15 Controlled torque curve

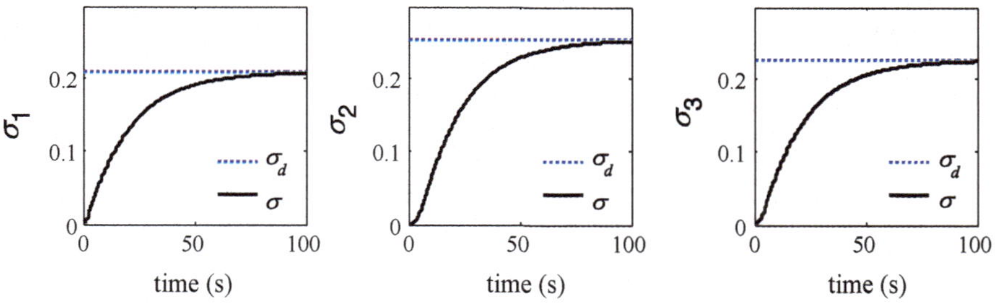

Fig. 6.16 MRP response of spacecraft body attitude

maneuver, which affects the attitude stability of the spacecraft. Therefore, this section carries out the research on the coupling control method for the flexible spacecraft. In the research of this section, the attitude control of the spacecraft adopts two control methods: robust H_∞ and adaptive sliding mode, which are respectively combined with the active vibration control method of flexible structures based on positive position feedback.

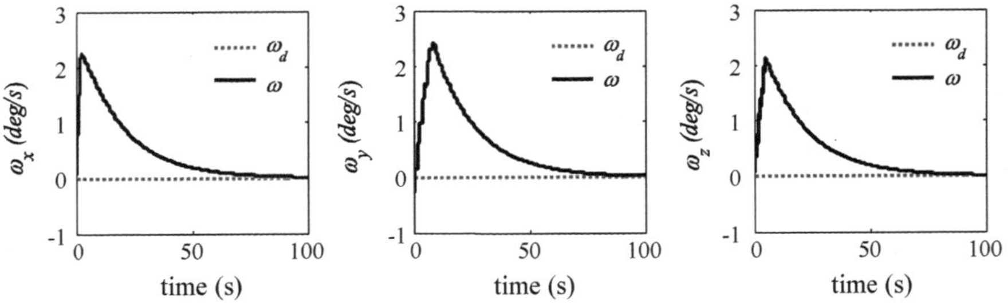

Fig. 6.17 Angular velocity response of spacecraft body attitude

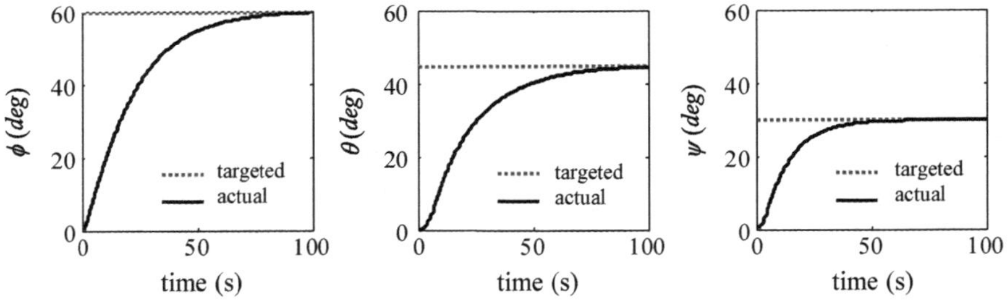

Fig. 6.18 Euler angle response of spacecraft body

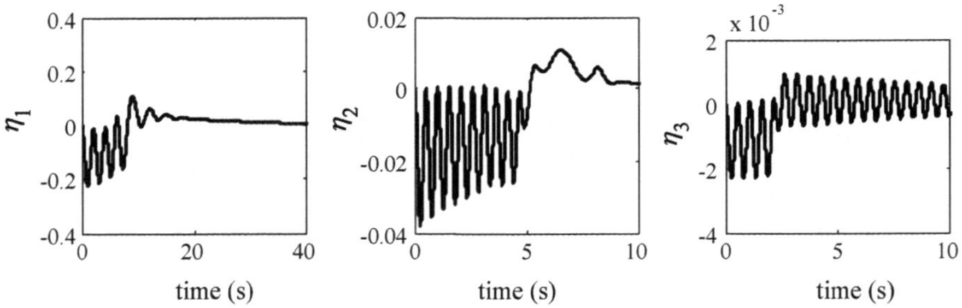

Fig. 6.19 Solar wing modal response

6.4.1 Principle of the Attitude Coupling Control Method

The coupling control system is composed of two subsystems, the attitude compensator subsystem and the vibration control subsystem, as shown in Fig. 6.21. The coupling controller receives signals of the elastic vibration and attitude. And then the controller calculates the output signals to vibration and attitude actuators.

The attitude control compensation subsystem mainly includes three parts: attitude sensor, attitude controller and attitude actuator. The attitude sensor measures the attitude parameters of the spacecraft relative to a certain reference frame and inputs

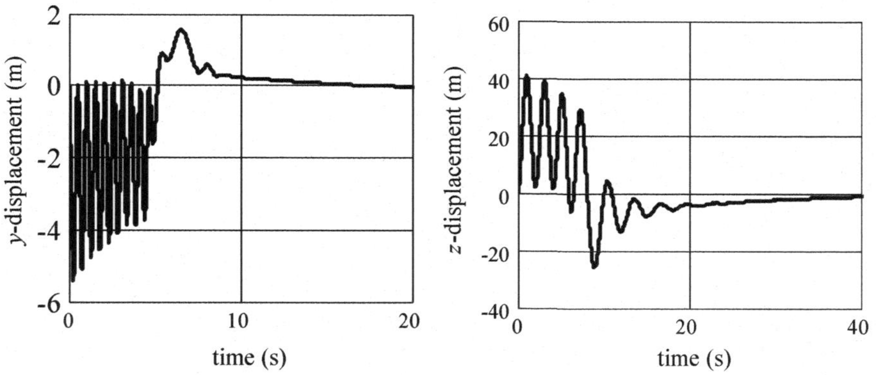

Fig. 6.20 Displacement response of solar wing tip

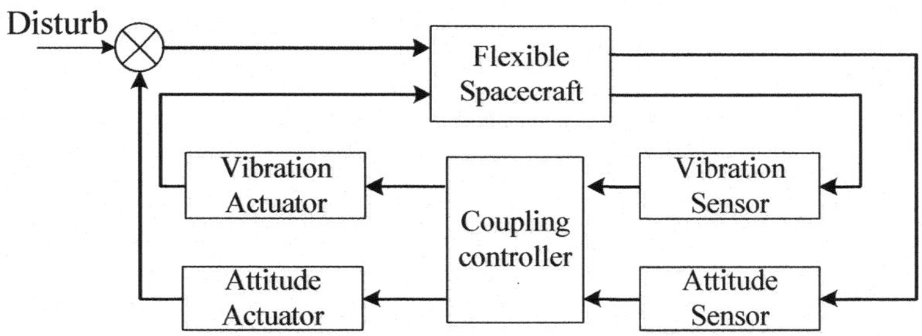

Fig. 6.21 Scheme of the coupling control strategy

them to the attitude controller. The required control force/torque signal is generated through the control algorithm, and the force/torque generated by the actuator acts on the spacecraft. The vibration control compensation subsystem mainly includes three parts: vibration sensor, controller and actuator. The vibration sensor measures the vibration and deformation information of the flexible structure, the controller calculates the required control signal, and the actuator generates a control force to act on the flexible structure to achieve the purpose of suppressing vibration. The attitude motion equation of the flexible spacecraft and the vibration equation of the flexible structure are coupled. When designing the attitude control law and vibration control law of a single subsystem, the influence of the other subsystem needs to be considered.

6.4.2 Attitude Coupling Controller Design

The elastic vibration of a flexible structure is suppressed by the force generated by bonded piezoelectric patches. Then the governing equation of the elastic vibration has the form

$$\boldsymbol{D}(t)^T\dot{\boldsymbol{\omega}} + \ddot{\boldsymbol{\eta}} + 2\boldsymbol{\xi}\boldsymbol{\Lambda}\dot{\boldsymbol{\eta}} + \boldsymbol{\Lambda}^2\boldsymbol{\eta} = \boldsymbol{B}_a\boldsymbol{V} \tag{6.58}$$

where $\boldsymbol{V}(m \times 1, m$ is the number of piezoelectric patches) denotes the control voltage, $\boldsymbol{B}_a$ is the distribution matrices of actuators.

In order to damp the elastic oscillations of flexible appendage, the PPF controller, which is robust to parametric uncertainties, is introduced. The PPF controller consists of a second-order controller. The second-order system is forced by the position response which is then feedback to give the force input to the structure. The governing equation of the second-order controller is given by

$$\ddot{\boldsymbol{\varepsilon}} + 2\boldsymbol{\xi}_c\boldsymbol{\Lambda}_c\dot{\boldsymbol{\varepsilon}} + \boldsymbol{\Lambda}_c^2\boldsymbol{\varepsilon} = \boldsymbol{\Lambda}_c^2\boldsymbol{B}_s^T\boldsymbol{\eta} \tag{6.59}$$

and

$$\boldsymbol{V} = \boldsymbol{G}\boldsymbol{\Lambda}^2\boldsymbol{\varepsilon} \tag{6.60}$$

where $\boldsymbol{G}$ is the feedback gain matrix of PPF, $\boldsymbol{\varepsilon}$ is the compensator state, $\boldsymbol{\xi}_c$ is damp of the compensator, $\boldsymbol{\Lambda}c$ is the compensator stiffness matrix, and $\boldsymbol{B}_s$ is the distribution matrices of sensors.

For closed-loop control systems that only consider the elastic vibration of flexible structures, the global dynamic equation with the PPF control law is given by

$$\begin{bmatrix} \ddot{\boldsymbol{\eta}} \\ \ddot{\boldsymbol{\varepsilon}} \end{bmatrix} + \begin{bmatrix} 2\boldsymbol{\xi}\boldsymbol{\Lambda} & 0 \\ 0 & 2\boldsymbol{\xi}_c\boldsymbol{\Lambda}_c \end{bmatrix}\begin{bmatrix} \dot{\boldsymbol{\eta}} \\ \dot{\boldsymbol{\varepsilon}} \end{bmatrix} + \begin{bmatrix} \boldsymbol{\Lambda}^2 & -\boldsymbol{B}_a\boldsymbol{G}\boldsymbol{\Lambda}^2 \\ -\boldsymbol{\Lambda}_c^2\boldsymbol{B}_s^T & \boldsymbol{\Lambda}_c^2 \end{bmatrix}\begin{bmatrix} \boldsymbol{\eta} \\ \boldsymbol{\varepsilon} \end{bmatrix} = \begin{bmatrix} -\boldsymbol{D}(t)^T\dot{\boldsymbol{\omega}} \\ 0 \end{bmatrix} \tag{6.61}$$

After the actuator and sensor distribution matrices are determined, the compensator parameters need to be determined, including the damping matrix, the stiffness matrix and the gain matrix. In order to effectively control structural vibration, active damping is required. In former literature, the stiffness matrix $\boldsymbol{\Lambda}_c$ is consistent with the natural frequencies of the target structure.

The closed-loop system in Eq. (6.61) is stable if only the following equation is satisfied

$$\boldsymbol{\Lambda}^2 - \boldsymbol{B}_a\boldsymbol{G}\boldsymbol{B}_s > 0 \tag{6.62}$$

The proof can be seen in the literature [22].

Then the nonlinear model in Eq. (6.58) for spacecraft motion including the PPF controller is summarized by

$$\begin{aligned} &\boldsymbol{J}(t)\dot{\boldsymbol{\omega}} + \boldsymbol{D}(t)\ddot{\boldsymbol{\eta}} + \boldsymbol{\omega} \times (\boldsymbol{J}\boldsymbol{\omega} + \boldsymbol{D}(t)\dot{\boldsymbol{\eta}}) = \boldsymbol{\tau} \\ &\boldsymbol{D}(t)^T\dot{\boldsymbol{\omega}} + \ddot{\boldsymbol{\eta}} + 2\boldsymbol{\xi}\boldsymbol{\Lambda}\dot{\boldsymbol{\eta}} + \boldsymbol{\Lambda}^2\boldsymbol{\eta} = \boldsymbol{B}_a\boldsymbol{G}\boldsymbol{\Lambda}^2\boldsymbol{\varepsilon} \end{aligned}$$

$$\ddot{\boldsymbol{\varepsilon}} + 2\boldsymbol{\xi}_c\boldsymbol{\Lambda}_c\dot{\boldsymbol{\varepsilon}} + \boldsymbol{\Lambda}_c^2\boldsymbol{\varepsilon} = \boldsymbol{\Lambda}_c^2\boldsymbol{B}_s^T\boldsymbol{\eta} \tag{6.63}$$

Define

$$\boldsymbol{q} = \begin{bmatrix} \boldsymbol{\eta} \\ \boldsymbol{\varepsilon} \end{bmatrix} \tag{6.64}$$

Inserting $\boldsymbol{q}$ into Eq. (6.63), and we obtain

$$\begin{aligned} &\boldsymbol{J}(t)\dot{\boldsymbol{\omega}} + \boldsymbol{D}_q(t)\ddot{\boldsymbol{q}} + \boldsymbol{\omega} \times \left(\boldsymbol{J}\boldsymbol{\omega} + \boldsymbol{D}_q(t)\dot{\boldsymbol{q}}\right) = \boldsymbol{\tau} \\ &\boldsymbol{D}_q(t)^T\dot{\boldsymbol{\omega}} + \ddot{\boldsymbol{q}} + 2\boldsymbol{\xi}_q\boldsymbol{\Lambda}_q\dot{\boldsymbol{q}} + \mathbf{K}_q\boldsymbol{q} = 0 \end{aligned} \tag{6.65}$$

where

$$\begin{aligned} &\boldsymbol{\xi}_q = \begin{bmatrix} \boldsymbol{\xi} & 0 \\ 0 & \boldsymbol{\xi}_c \end{bmatrix} && \boldsymbol{\Lambda}_q = \begin{bmatrix} \boldsymbol{\Lambda} & 0 \\ 0 & \boldsymbol{\Lambda}_c \end{bmatrix} \\ &\mathbf{K}_q = \begin{bmatrix} \boldsymbol{\Lambda}^2 & -\boldsymbol{B}_a\boldsymbol{G}\boldsymbol{\Lambda}^2 \\ -\boldsymbol{\Lambda}_c^2\boldsymbol{B}_s^T & \boldsymbol{\Lambda}_c^2 \end{bmatrix} && \boldsymbol{D}_q = [\boldsymbol{D}(t) \quad 0] \end{aligned} \tag{6.66}$$

Then the attitude control method is designed on the basis of the system shown in Eq. (6.65) to realize the coupled control of the flexible spacecraft. The design of the coupled controller using the robust H_∞ attitude control method is shown in Sect. 6.2.1. The design of the coupled controller adopting the adaptive sliding mode control method is shown in Sect. 6.3.1. Comparing Eq. (6.65) with Eq. (6.33), the controlled torque can be expressed as

$$\boldsymbol{\tau} = \boldsymbol{\omega} \times \boldsymbol{J}_{nom}\boldsymbol{\omega} + \boldsymbol{J}_{nom}\dot{\hat{\boldsymbol{\omega}}}(t) - \Lambda\boldsymbol{J}_{nom}\dot{\boldsymbol{G}}^{-1}(\sigma)\sigma_e(t) - \Lambda\boldsymbol{J}_{nom}\boldsymbol{G}^{-1}(\sigma)\dot{\sigma}_e(t) + \boldsymbol{\tau}' \tag{6.67}$$

The meaning of each parameter in the above equation is shown in Sect. 6.3.1.

6.4.3 Numerical Simulations with Robust Control and PPF[1]

The simulation example is consistent with Sect. 6.2.2.1. PZT layers of dimensions 85 × 14 × 0.3 mm are bonded to the fixed end of the plate to act as actuators. Properties of PZT are given in the Table 6.3. The stiffness and the mass of the piezoelectric layers are incorporated into the model and the structural damping is ignored in this case. The input voltage of the actuator is restricted to the range of [−500, 1500]V.

[1] J. Wang, J. Wu, W. Liu and H. Ji, "Coupling Attitude Control for Flexible Spacecraft with Rotating Structure," 2020 4th International Conference on Robotics and Automation Sciences (ICRAS), 2020, pp. 67–71.

Table 6.3 Typical properties of piezoelectric materials

Items	Parameters	Items	Parameters
Young's Modulus (GPa)	30.336	d_{33}(pC/N)	460
Poisson ratio	0.31	d_{31}(pC/N)	−210
Density (g/cm^3)	5.44	Voltage range (V)	[−1000, 1000]

The hub-plate system undergoes a 10 deg angle rest to rest slew, using the coupling control method. And the results are compared to the case that only the adaptive sliding mode control is utilized. Parameters of the PPF controller are set as $\boldsymbol{\Lambda}_c = \boldsymbol{\Lambda}$, $\xi_c = 0.4$. Parameters of the adaptive sliding mode controller are set as $\lambda = 1.0$, $k = 1.0$, $\epsilon = 0.001$. For simplicity, we only consider the first four flexible modes of the plate. And the modal damping ratios are all considered to be 0.01.

The results of robust control and coupled control are shown in Figs. 6.22, 6.23, 6.24, 6.25 and 6.26. Figure 6.22 shows the operating voltage of the piezoelectric sheet

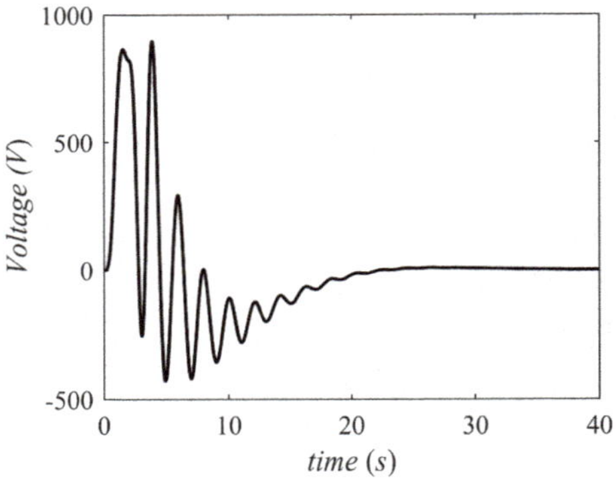

Fig. 6.22 Piezoelectric film actuation voltage curve

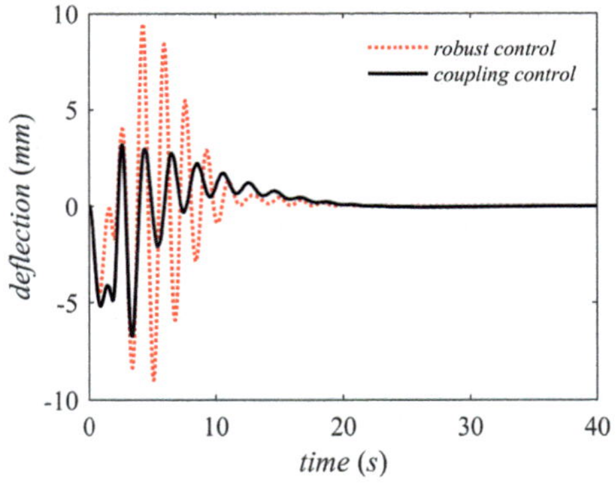

Fig. 6.23 Comparison of radial vibration response at the top of flexible plate

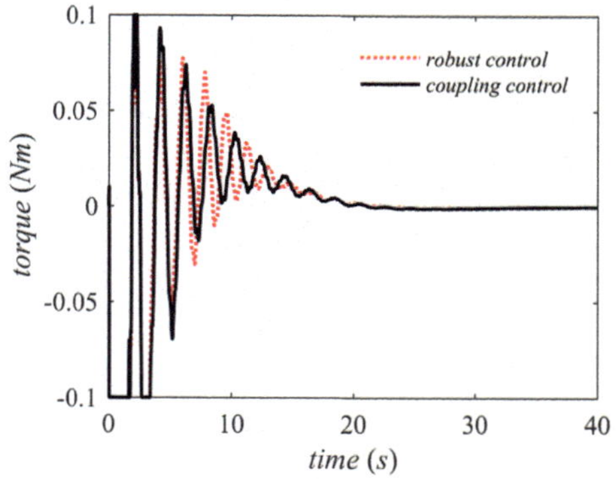

Fig. 6.24 Controlled torque input comparison

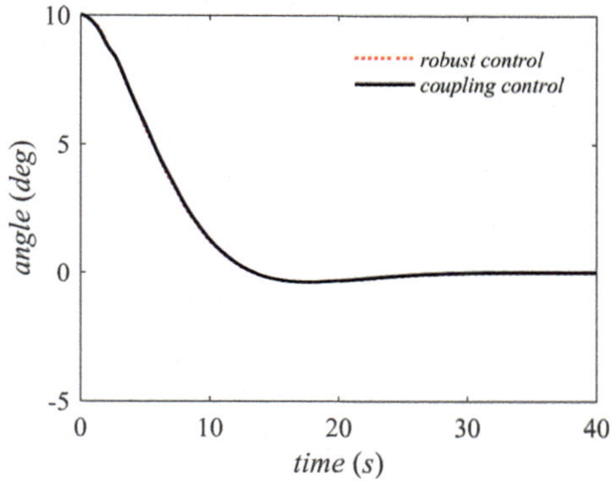

Fig. 6.25 Comparison of attitude angle response of center rigid body

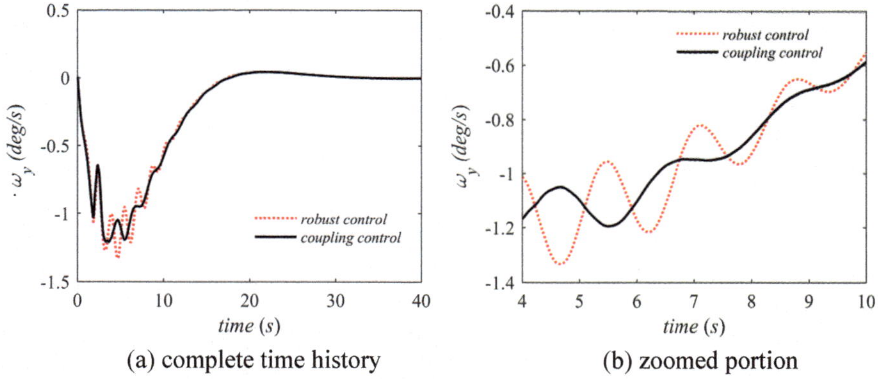

Fig. 6.26 Comparison of angular velocity response of central rigid body attitude

when the coupled controller is used. During the maneuvering process, the voltage varies within the range of [−429, 895]V.

Figure 6.23 shows the flexural vibration response of the top of the flexible plate. When the coupling controller is used, the amplitude of the vibration of the excited flexible plate is significantly reduced, and the jitter of the angular velocity of the central hub is correspondingly reduced (Fig. 6.26b). In the case of robust control, the elastic vibration amplitude of the flexible plate excited by the attitude maneuver is 18.4 mm while in the case of the coupling control the amplitude is 10.2 mm. At 15 s after the start of the control, the vibration amplitude is attenuated by 0.82 mm due to the damping and compensation of the controller.

In Fig. 6.25, the attitude angle responses in the case of robust control and coupled control are compared. At 25.6 s after the start of the attitude maneuver, the attitude angle reached the target value. It can be seen that after the PPF controller is applied to the flexible plate, the time of the maneuver is not substantially changed, and the trends of the controlled torque, the attitude angular velocity, and the attitude angle are consistent. Due to the attenuation of the elastic vibration of the flexible plate, amplitudes of jitter of the controlled torque and the attitude angular velocity during the maneuver are reduced, and the controlled torque curve is smoother.

The numerical simulation results show that compared with the attitude control alone, the coupling control method shortens the stability time of the spacecraft attitude and improves the pointing stability of the spacecraft.

6.4.4 Numerical Simulations with Adaptive Sliding Mode Control and PPF

The central hub performs a 60-degree attitude maneuver around the *y*-axis under the action of the adaptive sliding mode controller, and the PPF controller is used to attenuate the excited flexible plate elastic vibration. The parameters of the adaptive sliding mode controller are consistent with those in Sect. 6.3.2.1. The parameters of the PPF controller are $\boldsymbol{\Lambda}_c = \boldsymbol{\Lambda}, \quad \boldsymbol{\xi}_c = 0.4$.

Figures 6.27, 6.28, 6.29, 6.30, 6.31 and 6.32 show the dynamic response comparison of the system when only the adaptive sliding mode controller is used and the coupled control method is used. Figure 6.28 shows the actuation voltage of the piezoelectric sheet. It can be clearly seen from Figs. 6.31 and 6.32 that the elastic vibration of the flexible plate is rapidly attenuated under the action of the PPF controller. At $t = 10$ s after the start of the maneuver, when the PPF controller is not used, the elastic vibration amplitude of the flexible plate is 1.73 mm, while the vibration amplitude with the PPF controller is 0.14 mm.

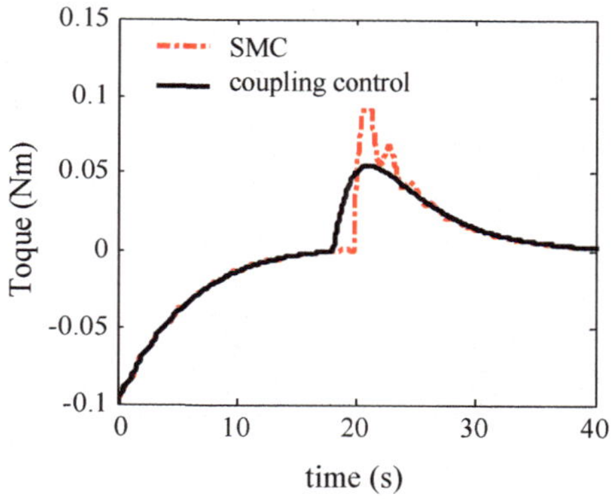

Fig. 6.27 Controlled torque input comparison

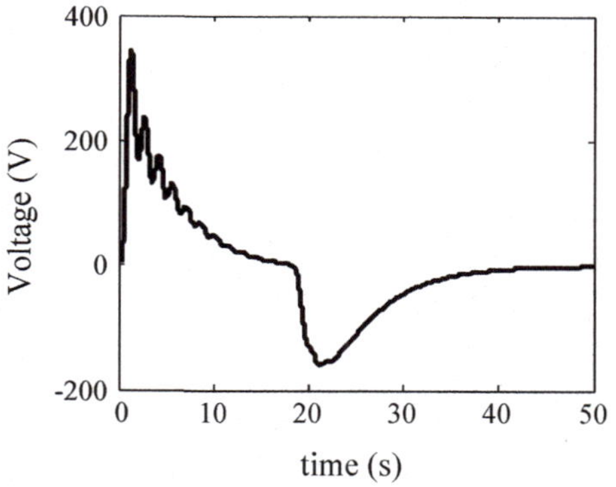

Fig. 6.28 Piezoelectric film actuation voltage curve

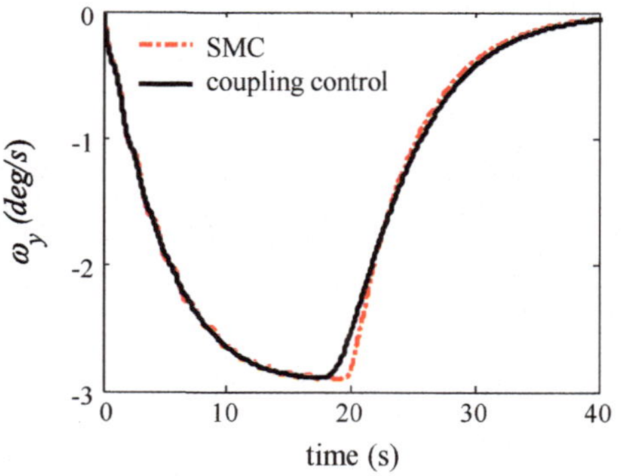

Fig. 6.29 Comparison of angular velocity response of central rigid body attitude

6.5 Summary

A coupling control strategy is designed in this chapter, which synthesizes an attitude controller and a vibration controller. Also, the method is robust to parametric uncertainties of the spacecraft model. Numerical simulations have provided satisfactory

Fig. 6.30 Comparison of attitude angle response of center rigid body

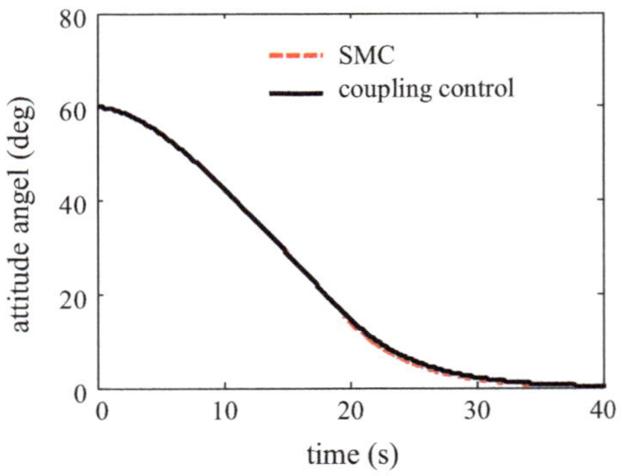

Fig. 6.31 Comparison of the first-order modal response of flexible plates

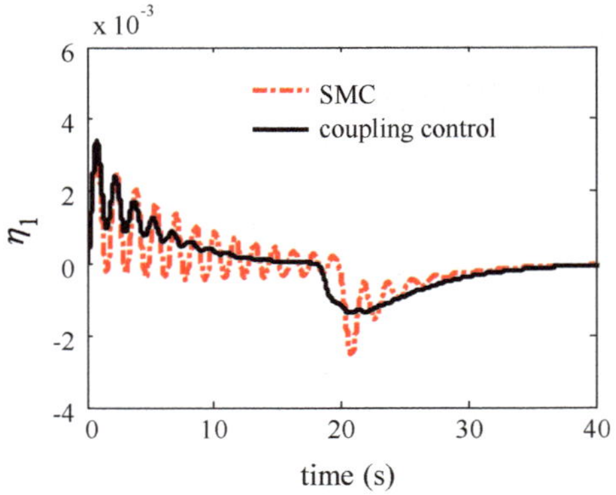

Fig. 6.32 Comparison of radial vibration response at the top of flexible plate

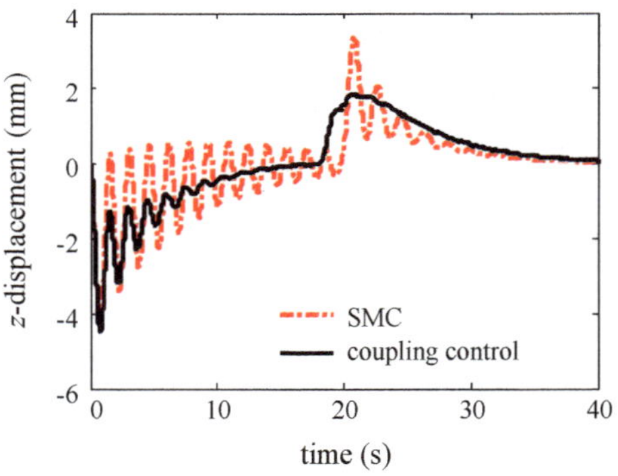

results in terms of vibration control and rigid motion. Simulations show that, the coupling control strategy reduces the stabilization time and improves the pointing accuracy of the platform. Hence, the method is significant for the flexible spacecraft to accomplish attitude maneuver missions.

References

1. Quinnt, R. D., & Meirovitch, L. (1986). Equations for the vibration of a slewing flexible spacecraft [J]. *AIAA*, 86–0906.
2. Kakad, Y. P. (1986). Dynamics and control of slew Manewer of large flexible spacecraft [J]. *AIAA, 86–47472*, 629–634.
3. Azadi, M., Eghtesad, M., & Fazelzadeh, S. et al. (2015). Dynamics and control of a smart flexible satellite moving in an orbit [J]. *Multibody System Dynamics*, 1–23.
4. Sales, T., Rade, D., & de Souza, L. (2013). Passive vibration control of flexible spacecraft using shunted piezoelectric transducers [J]. *Aerospace Science and Technology, 29*(1), 403–412.
5. Silverberg, L., & Baruh, H. (1988). Simultaneous maneuver and vibration suppression of flexible spacecraft [J]. *Applied Mathematical Modelling, 12*(6), 546–555.
6. Baruh, H. (2001). Control of the elastic motion of lightweight structures; proceedings of the *AIAA/ASME/ASCE/AHS/ASC Structures, Structural Dynamics, and Materials Conference and Exhibit*, 42 nd, Seattle, WA, F, 2001 [C].
7. Meirovitch, L., & Kwak, M. K. (1990). Dynamics and control of spacecraft with retargeting flexible antennas [J]. *J Guidance, 13*(2), 241–248.
8. Gennaro, S. D. (1998). Active vibration suppression in flexible spacecraft attitude tracking [J]. *Journal of Guidance, Control, and Dynamics, 21*(3), 400–408.
9. Grewal, A., & Modi, V. (1996). Robust attitude and vibration control of the space station [J]. *Acta Astronautica, 38*(3), 139–160.
10. Zhu, L., Liu, Y., & Wang, D. et al. (2008). Backstepping-based attitude maneuver control and active vibration reduction of flexible spacecraft. *Proceedings of the Control and Decision Conference*, 2008 CCDC 2008 Chinese, F, 2008 [C]. IEEE.
11. Azadi, M., Fazelzadeh, S., Eghtesad, M., et al. (2011). Vibration suppression and adaptive-robust control of a smart flexible satellite with three axes maneuvering [J]. *Acta Astronautica, 69*(5), 307–322.
12. Meiyu, C., & Shijie, X. (2010). Optimal attitude control of flexible spacecraft with minimum vibration. *Proceedings of the AIAA Guidance, Navigation and Control Conference*, AIAA-2010–8201, F, 2010 [C]. AIAA Toronto.
13. Shahravi, M., & Azimi, M. (2014). A comparative study for collocated and non-collocated sensor/actuator placement in vibration control of a maneuvering flexible satellite [J]. *Proceedings of the Institution of Mechanical Engineers, Part C: Journal of Mechanical Engineering Science*, 0954406214544182.
14. Qiu, Z. (2006). Active vibration control for coupling system of flexible structures and rigid body [J]. *Chinese Journal of Mechanical Engineering, 42*(11), 26–33.
15. Song, G., & Agrawal, B. N. (2001). Vibration suppression of flexible spacecraft during attitude control [J]. *Acta Astronautica, 49*(2), 73–83.
16. Hu, Q., & Ma, G. (2005). Active vibration suppression in flexible spacecraft attitude maneuver using variable structure control and input shaping technique. *Proceedings of the AIAA Guidance, Navigation, and Control Conference and Exhibit*, F, 2005 [C].
17. Hu, Q. (2012). Robust adaptive sliding mode attitude control and vibration damping of flexible spacecraft subject to unknown disturbance and uncertainty [J]. *Transactions of the Institute of Measurement and Control, 34*(4), 436–447.
18. Hu, Q., & Ma, G. (2005). Variable structure control and active vibration suppression of flexible spacecraft during attitude maneuver [J]. *Aerospace Science and Technology, 9*(4), 307–317.
19. Hu, Q., & Ma, G. (2005). Vibration suppression of flexible spacecraft during attitude maneuvers [J]. *Journal of Guidance, Control, and Dynamics, 28*(2), 377–380.
20. Hu, Q., & Xiao, B. (2011). Robust adaptive backstepping attitude stabilization and vibration reduction of flexible spacecraft subject to actuator saturation [J]. *Journal of Vibration and Control, 17*(11), 1657–1671.

21. Na, S., Tang, G.-A., & Chen, L.-F. (2014). Vibration reduction of flexible solar array during orbital maneuver [J]. *Aircraft Engineering and Aerospace Technology: An International Journal, 86*(2), 155–64.
22. Fanson, J. L., & Caughey, T. K. (1990). Positive position feedback control for large space structures [J]. *AIAA Journal, 28*(4), 717–724.

Chapter 7
Rigid-Flexible Coupling Control Experiments

7.1 Introduction

In Chap. 5, an optimization method for variable amplitude input shaping attitude maneuver was proposed. And in Chap. 6, a coupled control method of attitude control and flexible structure vibration control was proposed for attitude tracking of flexible spacecraft with time-varying parameters. In order to verify the effectiveness of the above methods, this chapter builds up a flexible spacecraft attitude control testbed and conducts spacecraft attitude control experiments. The test results are compared with the simulation results to evaluate the performance of the methods.

7.2 Attitude Control Testbed for Flexible Spacecraft

In order to verify the effectiveness of attitude control methods proposed in previous chapters in the ground environment, a flexible spacecraft attitude control testbed is established in this section. This section introduces the concept and structure of the testbed.

The testbed consists of a single-axis air bearing table, an experimental flexible spacecraft, an attitude sensor, an attitude actuator, a vibration measurement subsystem, a vibration actuator, and a real-time control system. The attitude sensor adopts a fiber optic gyroscope and the attitude actuator utilizes a reaction wheel. A strain gauge and piezoelectric sheets act as the measuring device and actuators, respectively. The schematic diagram of the test system is shown in Fig. 7.1.

Figure 7.2 presents the photograph of the attitude control testbed for flexible spacecraft, which consists of a hub and a fixed aluminum rectangular plate. The hub represents the platform of a flexible spacecraft. And an optical fiber gyro and a

© The Author(s), under exclusive license to Springer Nature Singapore Pte Ltd. 2022

J. Wang and D.-X. Li, *Rigid-Flexible Coupling Dynamics and Control of Flexible Spacecraft with Time-Varying Parameters*,
https://doi.org/10.1007/978-981-16-5097-0_7

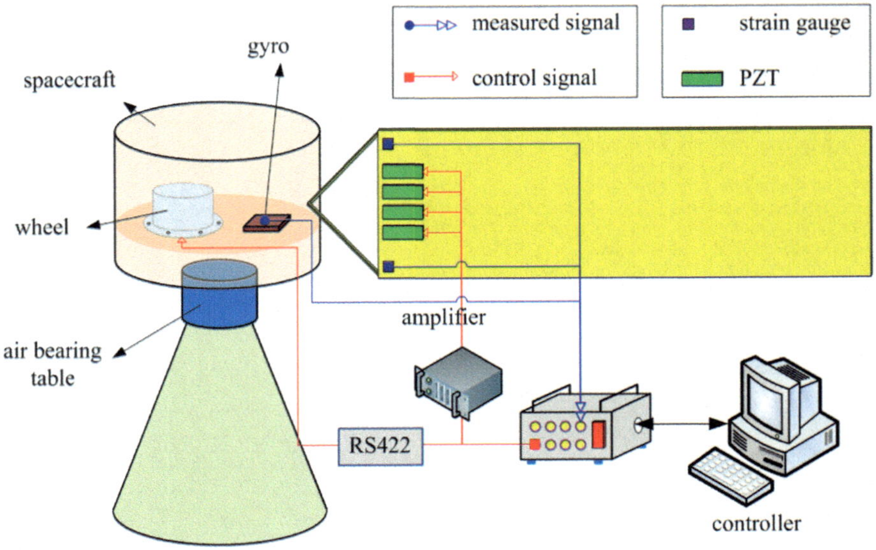

Fig. 7.1 Schematic diagram of flexible spacecraft coupling control system

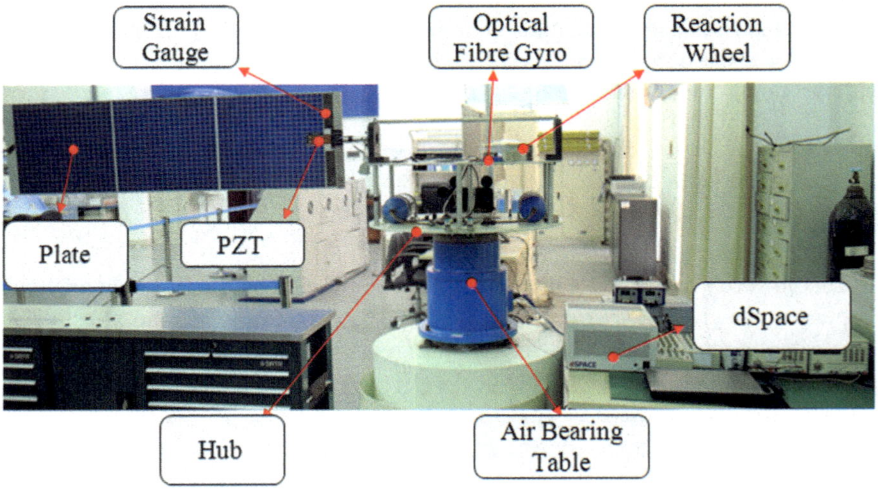

Fig. 7.2 Picture of the attitude control testbed of flexible spacecraft

reaction wheel are mounted on the hub. The gyro senses the angular velocity of the hub and the wheel supplies control moment to the system. The maximum control moment that the wheel can provide is $\pm$ 0.1 Nm. The plate is designed to simulate the behavior of flexible appendages, such as solar wings. The plate is bonded with a strain gauge and collocated PZT patches. They serve as sensors and actuators to sense or control the plate. The hub-plate system is mounted on a single axis air bearing table, which is used to provide similar conditions of low friction and the weightlessness of outer space. The testbed can perform single axis rotation to test attitude control algorithms and vibration control methods. The parameters of each instrument are shown in Table 7.1.

The experimental flexible spacecraft includes a platform and a flexible plate which represents the flexible solar wing. The size of the platform is $\Phi 800 \times 700$ mm, the mass is about 58 kg, and the moment of inertia is about 7.33 kg·m^2. The flexible plate is made of aluminum, whose size is $1.5 \times 0.4 \times 1.4$ mm. The plate is installed on the side of the platform through a yoke. The moment of inertia of the entire spacecraft around the axis of rotation is about 10.00 kg·m^2.

The testbed is a hardware in loop simulation platform, which is capable of validating coupling control strategies. Meanwhile, this system can be utilized to evaluate performances of vibration control methods for appendages and maneuver trajectories for flexible spacecraft. The testbed is realized using a simulation computer to provide

Table 7.1 List of experimental system instruments and technical parameters

Serial number	Instrument	Type	Technical index	Application
1	Air floatation table	–	Single axis	Simulation of frictionless, microgravity environment
2	Fiber optic gyroscope	VG910	Scale factor: 46.9 mV/deg/s Scale factor stability: 0.1%	Attitude sensor
3	Reaction wheel	4Nms	Working speed range: ±5000 r/min Maximum reaction torque: 0.1 Nm Speed control accuracy:better than ±1 rpm	Attitude actuator
4	Strain gauges	BF350-3AA	Resistance: 350.0 ± 0.1Ω Sensitivity factor: 2.15 ± 1%	Strain measuring device
5	Industrial amplifier	MP30	–	Amplify the strain signal
6	Piezoelectric sheet	M-8514-P1	See Table 6.3	Actuator
7	Power amplifier	E-500.621	Output voltage range: 100 ~ 240 V	Drive piezoelectric sheet
8	dSpace	–	Control panel: 1005 A/D: DS2002 D/A: DS2103	Controller

(a) (b) (c) (d)

Fig. 7.3 Some equipment diagrams: **(a)** fiber optic gyroscope; **(b)** reaction wheel; **(c)** piezoelectric sheet; **(d)** power amplifier

real-time calculation and multi-channel A/D, D/A connector panels for data acquisition and control output. When conducting experiments, analog signals generated by the gyro, strain gauge and PZT patches are fed back to the simulation computer through the A/D panel. The reaction torque value for the wheel and the control voltage values for PZT patches are determined by the coupling control strategy. The torque value is converted to data flow and the data is transmitted to the wheel through serial port. The control voltage values are transmitted to an amplifier through the D/A panel and then drive the PZT patches. To eliminate the influence of noise and high modes, low pass filters are utilized to smooth the outputs of sensors (Fig. 7.3).

7.3 Attitude Maneuver Experiment

7.3.1 Experimental Method

The central hub performs attitude maneuver with bang-bang controlled torque and the optimized controlled torque according to the variable amplitude input shaping strategy. Responses are compared to evaluate the proposed method.

The central hub starts attitude maneuver at $t = 5$ s. For the bang-bang maneuver, the controlled torque is shown in Fig. 7.4a. In the time interval $t = [5, 14.5$ s$]$, the input torque is 0.08 Nm, and in the interval $t = [14.5, 24$ s$]$, the input torque is -0.08 Nm. The input torque after input shaping is shown in Fig. 7.4b.

7.3.2 Data Analysis

The fiber optic gyroscope (FOG) and the strain gauge measure the attitude angular velocity of the central hub and the micro strain at the root of the flexible board, respectively, as shown in Figs. 7.5 and 7.6. The measured signals contain burrs due to noise and other factors, and are filtered by a low-pass filter. The filtered signals have better smoothness and are used as the input of the controller.

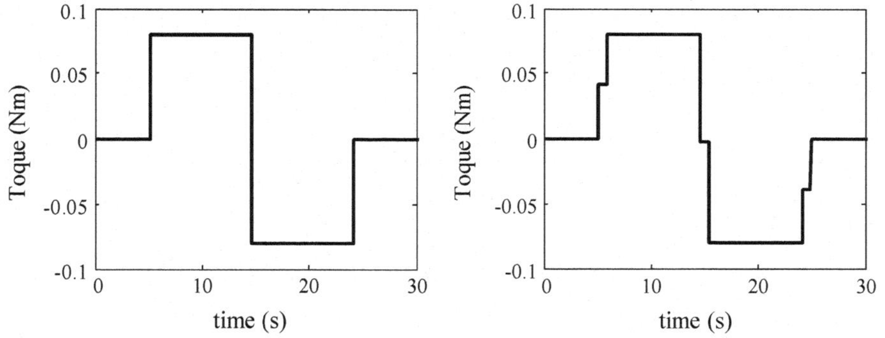

(a) bang-bang (b) Shaped input

Fig. 7.4 Controlled torque for central hub

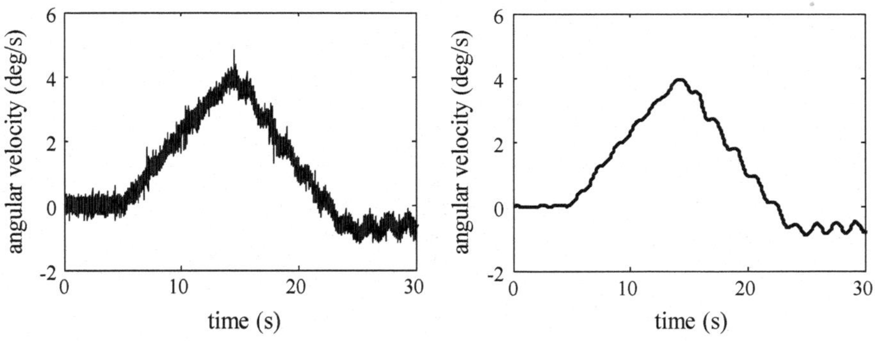

(a) Before filtering (b) After filtering

Fig. 7.5 Attitude angular velocity data measured by FOG

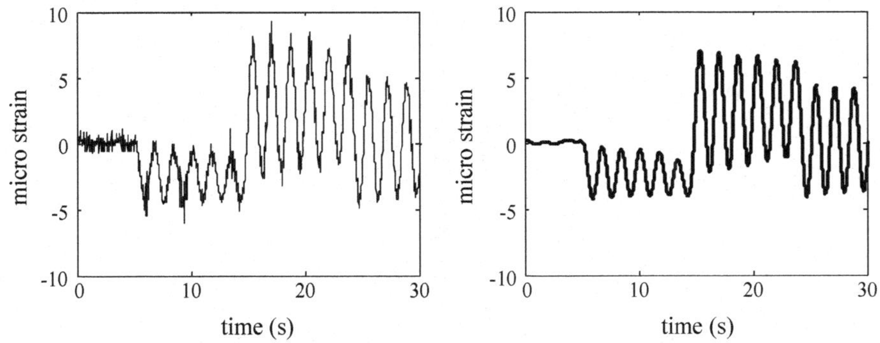

(a) Before filtering (b) After filtering

Fig. 7.6 Strain data measured by the strain gauge set

Figures 7.7, 7.8, 7.9 and 7.10 show dynamic responses of the system. Figure 7.7 shows the actual torque provided by the reaction wheel. Figure 7.8 shows the attitude angular velocity of the central hub. The central hub first accelerates uniformly and

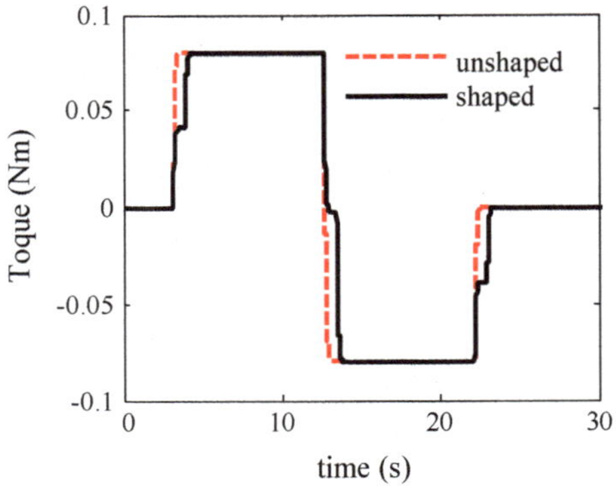

Fig. 7.7 Controlled torque of reaction wheel

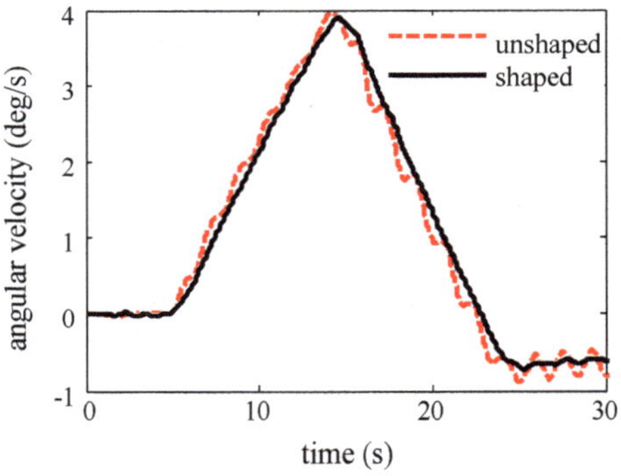

Fig. 7.8 Angular velocity of central hub

Fig. 7.9 Attitude angle of central hub

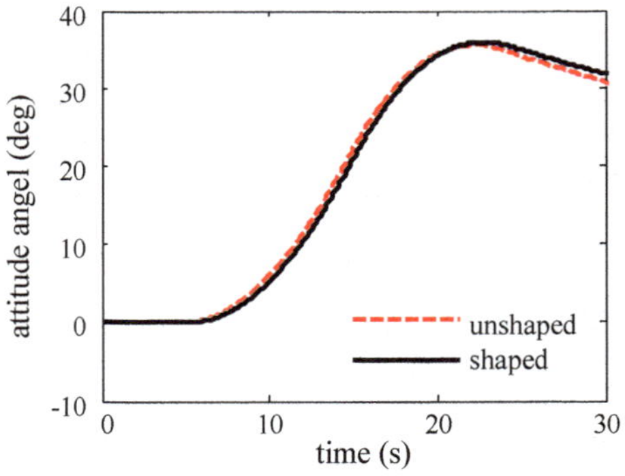

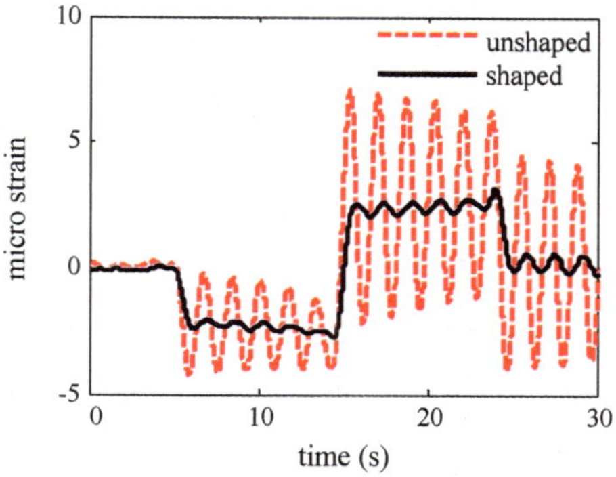

Fig. 7.10 Strain on the root of plate

then decelerates uniformly. Theoretically, the attitude angular velocity of the central hub is zero after the maneuver is completed. However, because of atmospheric resistance, the angular velocity is less than zero. In addition, under the bang-bang torque, the angular velocity exhibits obvious jitter during and after the maneuver, and the jitter amplitude after the maneuver is about 0.36 °/s. And the amplitude of jitter under the shaped torque is 0.055 °/s.

Figure 7.9 shows the response of attitude angle of the central hub. The response curves of the two maneuvering methods are basically consistent. The attitude angle is maneuvered from 0 to 35.97 °. Figure 7.10 shows the strain response on the root of the flexible plate. It can be seen that the stimulated elastic vibration of the flexible plate under shaped torque is significantly reduced. After maneuvering, the elastic vibration amplitude is reduced by an order of magnitude (from 8.78 to 0.73 microstrain).

7.4 Attitude Coupling Control Experiment[1]

7.4.1 Experimental Method

In this experiment, the testbed is commanded to perform a 60 ° slew, with respect to the initial attitude. For comparative purposes, two different cases are conducted: (1) slew using the adaptive sliding mode control method, (2) slew using the coupling control strategy, i.e., the adaptive sliding mode control for attitude and the PPF control for elastic vibration. Parameters of the PPF controller are set to $\boldsymbol{\Lambda}_c = \boldsymbol{\Lambda}$, $\xi_c = 0.4$. Parameters of the adaptive sliding mode controller are set to $\lambda = 0.2$, $k = 0.1$, $\epsilon = 0.05$.

[1] Jie Wang, Dongxu Li, Experiments study on attitude coupling control method for flexible spacecraft, Acta Astronautica, 2018, 147(1), 393–402.

7.4.2 Data Analysis

(1) **Performance analysis of coupling control**

Figures 7.11, 7.12, 7.13, 7.14, 7.15 and 7.16 presents the experimental results. Figure 7.12 presents the torque applied to the hub. Though the time histories are close to each other, the zoomed portion illustrated in Fig. 7.12b shows that there is indeed an influence of the plate elastic motion on the controlled torque. With only the adaptive sliding mode control is applied, there exists jitter for the controlled torque. It is clear that, with the PPF control method, the excited vibration of the plate is suppressed rapidly, as shown in Fig. 7.15. Such influence, indicated by oscillations, is reduced by the PPF controller. Similarly, Fig. 7.13a depicts the time histories of the angular velocities of the hub. The corresponding zoomed portion is depicted in Fig. 7.13b. One can see once again the influence of the flexibility of the plate on the rigid-motion responses. This reflects advantages of the coupling control strategy.

Figures 7.13 and 7.14 show the results in terms of attitude behavior. Also, it can be seen that the goals established for attitude control are satisfactorily achieved, namely, a steady 60 ° position is reached in nearly 40 s of control action.

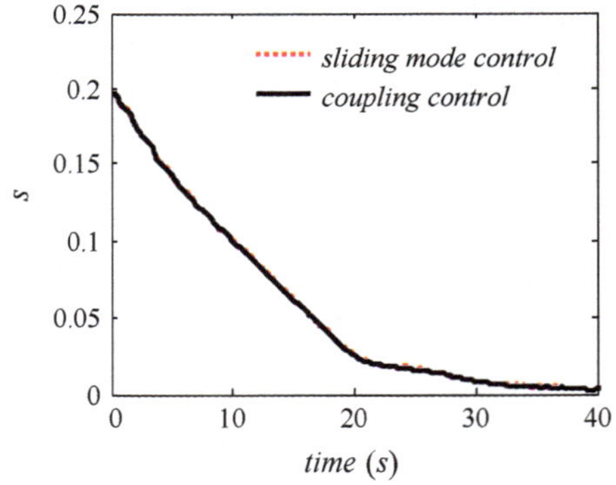

Fig. 7.11 The sliding manifold

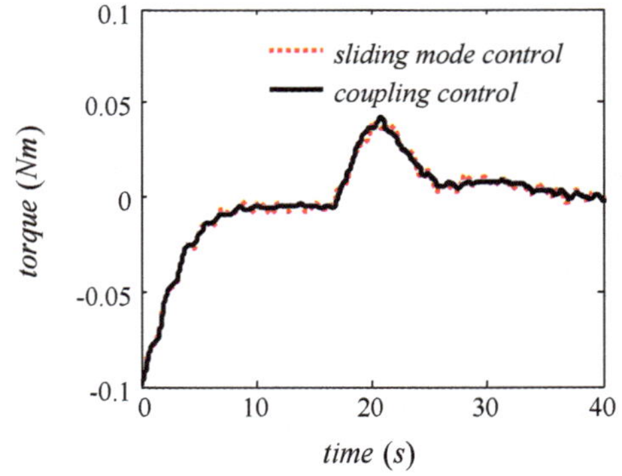

Fig. 7.12 Controlled torque of a rest to rest slew

Fig. 7.13 Angular velocity of the hub

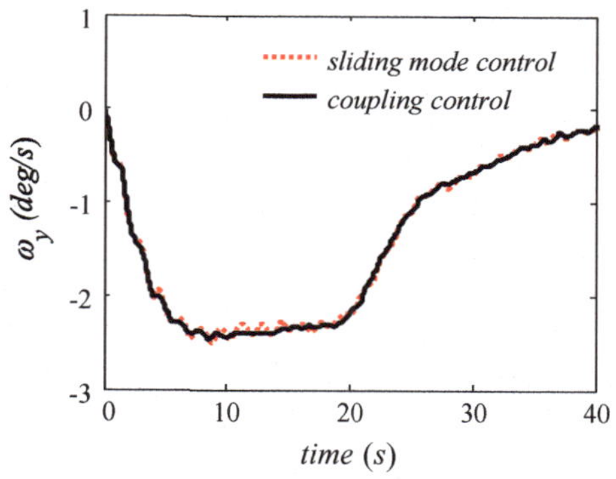

Fig. 7.14 Euler angle of the Hub

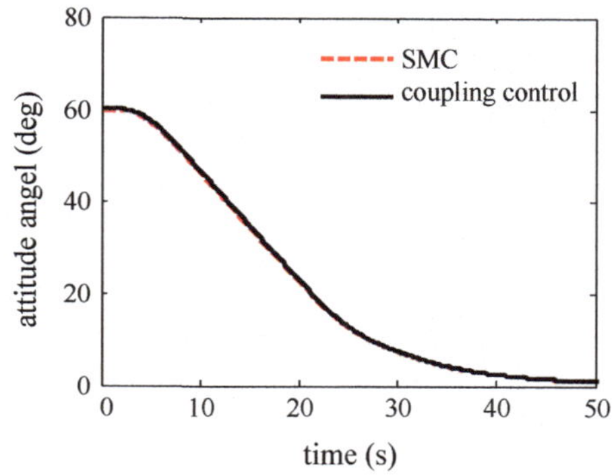

Fig. 7.15 Strain on root of the plate

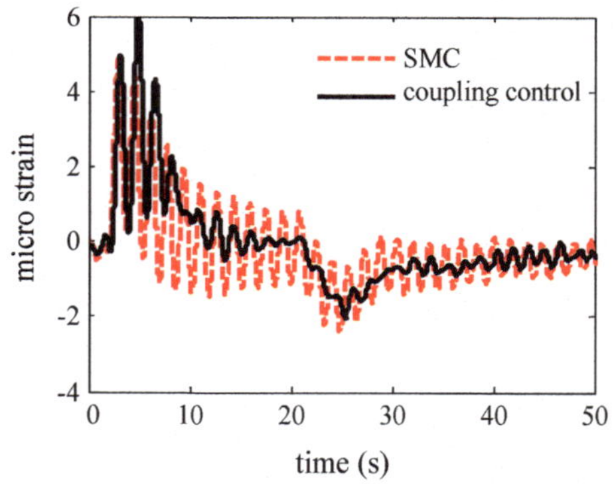

(2) **Parameter analysis of the controller**

In order to analyze the effect of the controller parameter on transient responses, three simulation cases with $\lambda = 0.2$, $\lambda = 0.1$ and $\lambda = 0.05$ are compared. And the results are shown in Figs. 7.17, 7.18, 7.19, 7.20, 7.21 and 7.22. When λ is increased, the time when states of the system move in the sliding surface boundary is delayed.

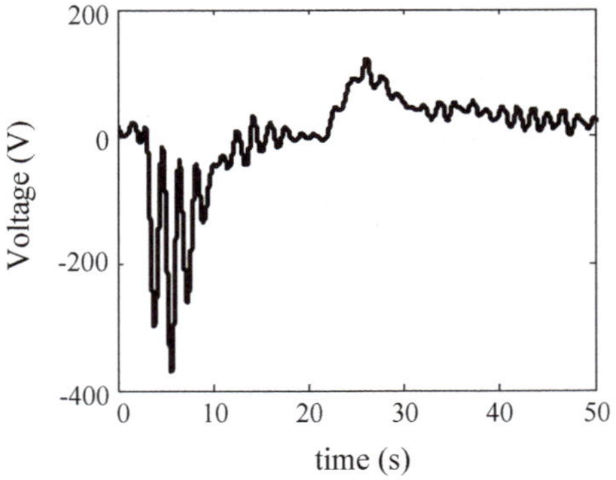

Fig. 7.16 Piezoelectric control voltage

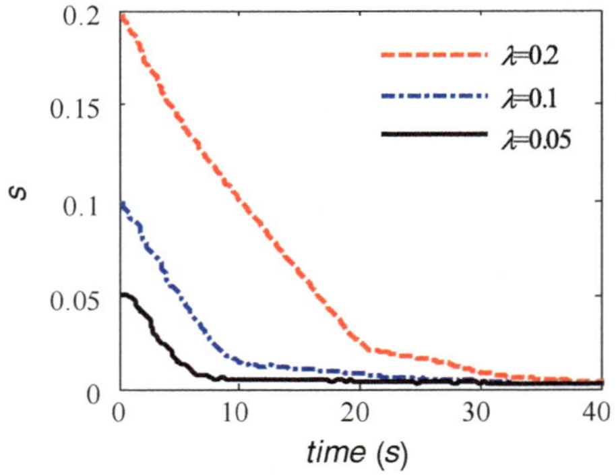

Fig. 7.17 The sliding manifold comparison with different parameters

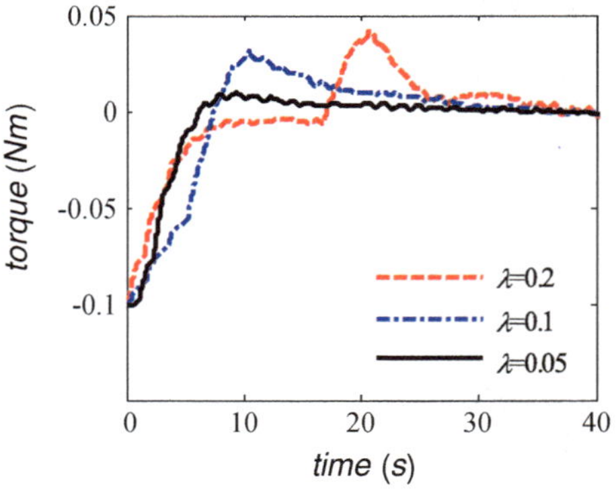

Fig. 7.18 Controlled torque comparison with different parameters

Meanwhile, states of the system converge toward the targeted value with a faster speed. The maximum amplitudes of the vibration of the flexible plate excited in the three cases are approximately equal.

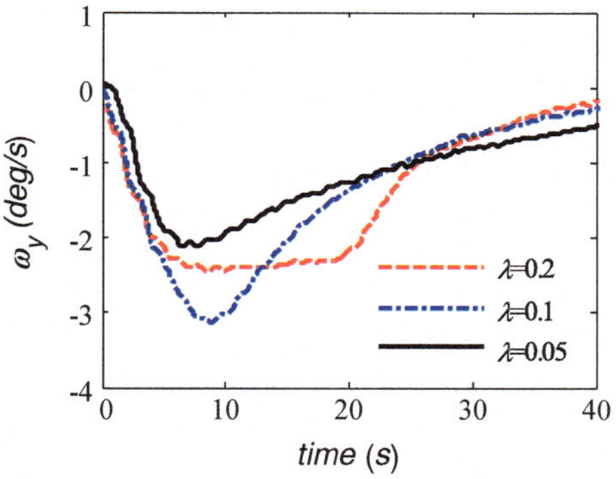

Fig. 7.19 Angular velocity comparison with different parameters

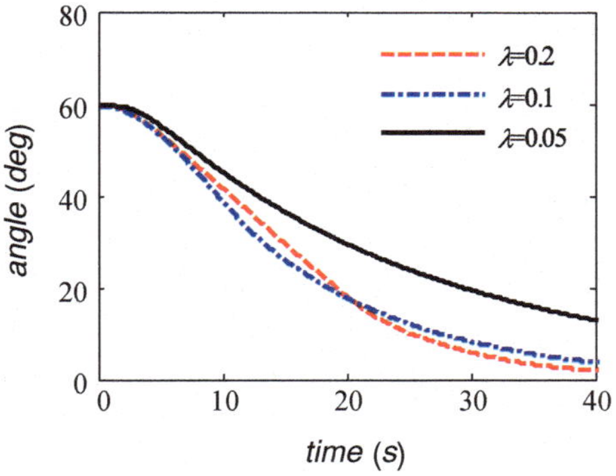

Fig. 7.20 Euler angle comparison with different parameters

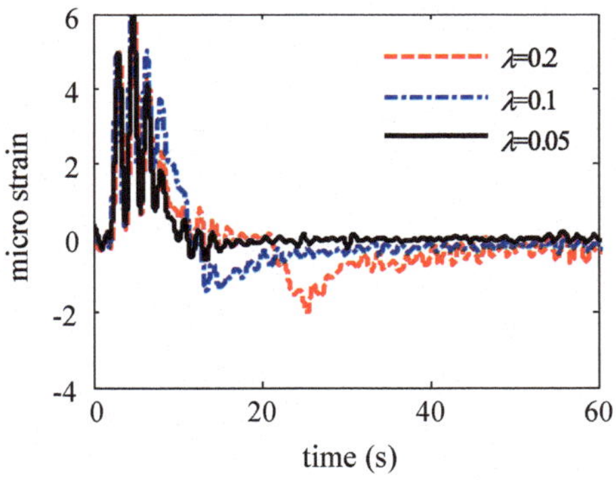

Fig. 7.21 Strain comparison on root of the plate

7.4.3 Comparison with Simulation Results

Figures 7.23, 7.24, 7.25, 7.26, 7.27 and 7.28 compare the experimental results with simulation results. From it is possible to conclude that the experimental maneuver trajectory of the hub is closely followed with the numerical result. Similarly,

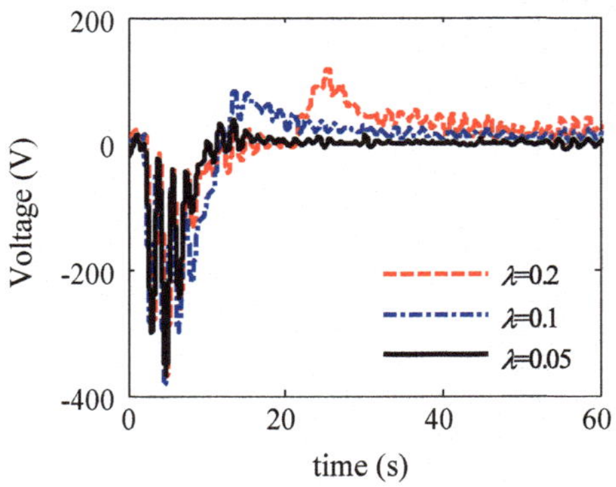

Fig. 7.22 Piezoelectric control voltage comparison

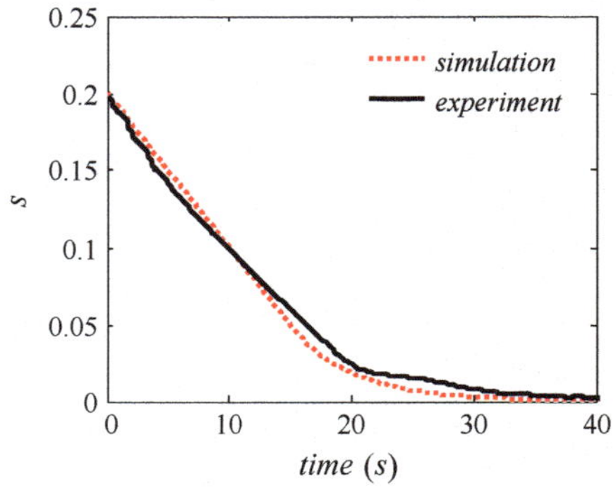

Fig. 7.23 The sliding manifold

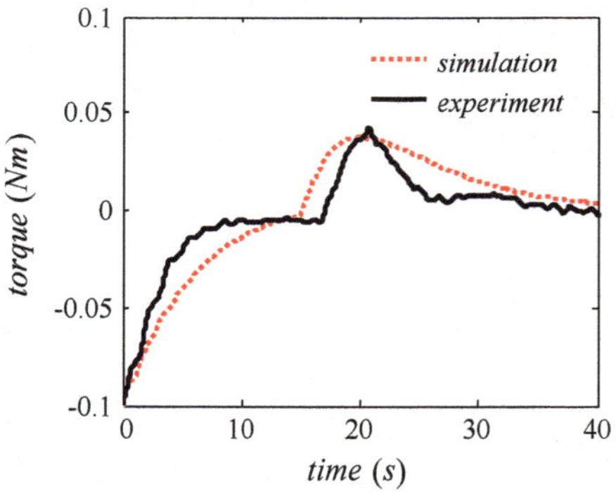

Fig. 7.24 Controlled torques

unwanted elastic oscillations presented clearly in the flexible plate. However, the oscillation in simulation decays faster than that in experiment. The mathematic model needs further modification to accommodate the experiment.

Fig. 7.25 Angular velocities of the Hub

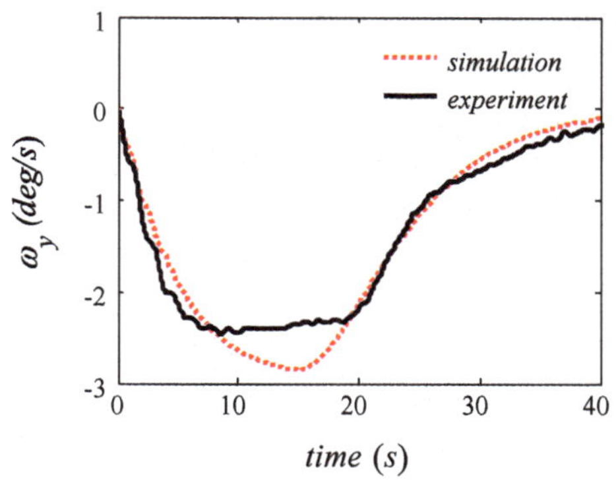

Fig. 7.26 Euler angle of the Hub

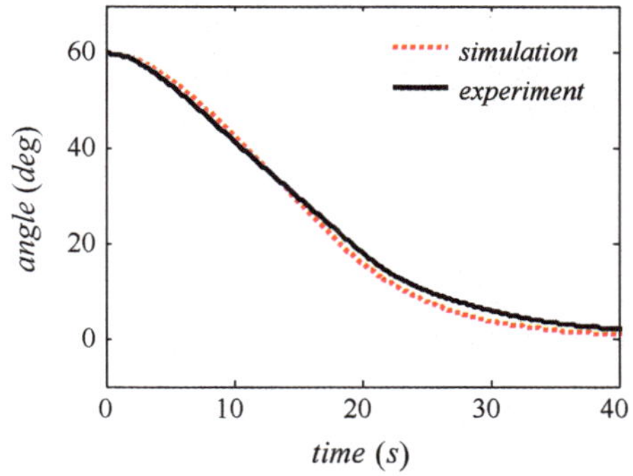

Fig. 7.27 Strain on the root of plate

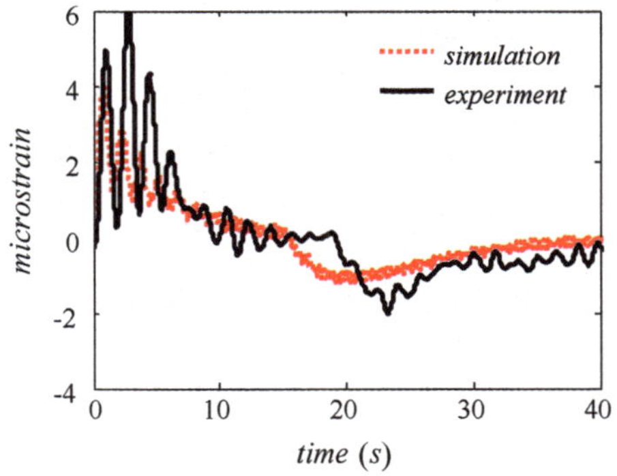

7.5 Summary

In this chapter, a flexible spacecraft attitude control testbed has been built and attitude control experiments have been carried out. Experimental results show that the maneuver path planning method based on input shaping can suppress the

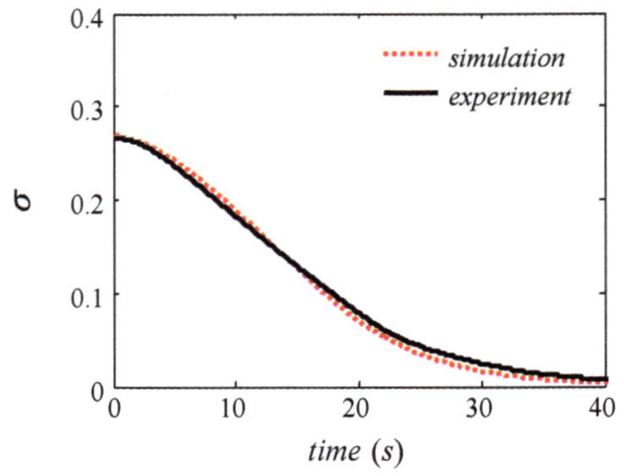

Fig. 7.28 Comparison of MRP

residual vibration of the flexible appendage and improve the attitude accuracy of the spacecraft. And the coupling control method is superior to the attitude control method.

Chapter 8
Future

High pointing accuracy and stabilization are significant for spacecraft to carry Earth observing, laser communication and space exploration missions. However, when the spacecraft undergoes large angle maneuver, the excited elastic oscillation of flexible appendages, for instance, solar wing and onboard antenna, would downgrade the performance of the spacecraft platform. Therefore, research on dynamic modeling and control of flexible spacecraft with time-varying parameters is of great significance. It is recommended to further deepen the research based on the work in this text. The dynamics model of flexible spacecraft can consider the influence of non-linear factors such as liquid sloshing, fuel consumption, friction, etc., to establish a more refined system dynamics model. In addition, feedback controllers can be considered in the variable amplitude input shaping maneuver method, such as PID control, linear state feedback, sliding mode control, robust control, intelligent control algorithms, etc. to transform the attitude maneuvering process into a closed-loop control system. Then the control input can be adjusted according to the real-time feedback of the system, and the robustness of the system would be enhanced. The variable-amplitude input shaping maneuvering method designed in this book can be improved from an offline system to an online system to enhance the adaptability of the method. To develop diverse coupling control methods, and further analyze the stabilities of coupling control methods. Try to establish a ground test system of flexible spacecraft which can perform three-axis attitude maneuvers to verify the effectiveness of the proposed methods.

© The Author(s), under exclusive license to Springer Nature Singapore Pte Ltd. 2022

J. Wang and D.-X. Li, *Rigid-Flexible Coupling Dynamics and Control of Flexible Spacecraft with Time-Varying Parameters*,
https://doi.org/10.1007/978-981-16-5097-0_8